Semi-inverse Method in Nonlinear Problems of Axisymmetric Shells Forming

Semi-inverse Method in Nonlinear Problems of Axisymmetric Shells Forming

ANATOLY S. YUDIN
Southern Federal University, Russia

DMITRY V. SHCHITOV
North-Caucasus Federal University, Russia

 **World Scientific**

NEW JERSEY · LONDON · SINGAPORE · BEIJING · SHANGHAI · HONG KONG · TAIPEI · CHENNAI · TOKYO

Published by

World Scientific Publishing Europe Ltd.

57 Shelton Street, Covent Garden, London WC2H 9HE

Head office: 5 Toh Tuck Link, Singapore 596224

USA office: 27 Warren Street, Suite 401-402, Hackensack, NJ 07601

Library of Congress Cataloging-in-Publication Data

Names: Yudin, Anatoly S., author. | Shchitov, Dmitry V., author.

Title: Semi-inverse method in nonlinear problems of axisymmetric shells forming /
    Anatoly S. Yudin, Southern Federal University, Russia, Dmitry V. Shchitov,
    North-Caucasus Federal University, Russia.

Description: London ; Singapore ; Hackensack, NJ : World Scientific, 2021. |
    Includes bibliographical references and index.

Identifiers: LCCN 2020051746 | ISBN 9781786349811 (hardcover) |
    ISBN 9781786349828 (ebook for institutions) | ISBN 9781786349835 (ebook for individuals)

Subjects: LCSH: Thin-walled structures--Mathematical models. | Elastic plates and shells--
    Mathematical models. | Nonlinear functional analysis. | Functions, Inverse.

Classification: LCC TA660.T5 Y83 2021 | DDC 624.1/776201515357--dc23

LC record available at https://lccn.loc.gov/2020051746

British Library Cataloguing-in-Publication Data

A catalogue record for this book is available from the British Library.

For any available supplementary material, please visit
https://www.worldscientific.com/worldscibooks/10.1142/Q0289#t=suppl

Desk Editors: Balamurugan Rajendran/Michael Beale/Shi Ying Koe

Typeset by Stallion Press
Email: enquiries@stallionpress.com

Printed in Singapore

Centenary dedicated to academician
Iosif Izrailevich Vorovich

About This Book

The monograph deals with the problems of large deformations of shells of rotation. The nonlinearity of problems includes both geometric and physical components. The geometric aspects of the shape change suggest the presence of large elongations and the need to take into account changes in the shell metric. The material of the shells are considered to be steel-type metals. Therefore, their physical properties are modeled on the basis of nonlinear diagrams of the material, which set the stress intensity as a function of logarithmic deformations. Loading is considered to be active, so that the process of plastic drawing can be interpreted as nonlinearly elastic. The used mathematical model is based on the equilibrium equations of E. Reissner, generalized for large elongation, and the ratio of the Davis–Nadai, for the construction of physical relations for the shell. The developed nonlinear models and solution methods are aimed at applications in some fields of technology such as the creation of high-precision membrane devices used to protect structures from destruction by overpressure, flat jacks used for lifting and leveling heavy structures, optimization of tank designs for the transport of liquid cargo, etc.

This book is recommended for researchers, graduate students, and senior students of physical and mathematical specialties, as well as scientific and technical workers dealing with thin-walled structures.

Preface

June 21, 2020, marked the 100th anniversary of the birth of Russian academician I.I. Vorovich. One of the authors was lucky enough to communicate and work closely with him for 45 years.

I.I. Vorovich entered the Department of Mechanics at Moscow State University in 1937. At the beginning of the Great Patriotic War, he was drafted into the Red Army and sent to study at the Air Force Engineering Academy. Having mastered the methods of high theory, he also received technical training and harsh practice on front-line airfields.

Therefore, the scientific school of academician I.I. Vorovich is characterized by work on both theoretical issues and applied tasks. The Institute of Mechanics and Applied mathematics, which he led as director, included both theoretical departments and experimental laboratories.

Following the traditions of the school of I.I. Vorovich, the book presents the development of mathematical models of shell theory and their application to the modeling of some technical objects. The designs and tasks that are solved for them appeared as a result of communication with engineers. The reader can get acquainted with flapping membranes used in security systems, problems of containers for transporting liquid dangerous goods, original shell-type flat jacks, seals in the form of hydraulic connection shells, etc.

The authors were not able to solve the problem of forming domed shells from round plates for a long time. Analysis of the results of experiments and the use of the semi-inverse method made it possible to achieve success. This method was further developed and applied to problems of changing the shape and to other shells. Therefore, the title of the book focuses on this approach.

The authors are grateful to Tatyana Shchitova for her help in editing the text of the book.

Research was financially supported by Southern Federal University, grant No. VnGr/2020-04-IM (Ministry of Science and Higher Education of the Russian Federation).

About the Authors

A.S. Yudin is Doctor of Physical and Mathematical Sciences and Chief Researcher at the Institute of Mathematics, Mechanics and Computer Science of the Southern Federal University. He is Adviser to the Rostov branch of the Russian Academy of Engineering. He was awarded the badge of "Honorary worker of higher professional education of the Russian Federation". He graduated from the Faculty of Mechanics and Mathematics of Rostov State University. He worked his way up from a Junior Researcher to Head of the Department of Thin-walled Structures at the Institute of Mechanics and Applied Mathematics. He led many scientific projects. He worked as a Professor at the Department of Construction Mechanics. He was a Member of the Academic Councils, and was also a Member of the Russian Acoustic Society. His areas of research include the following: shell theory, mathematical modeling, linear and nonlinear problems, stability of shells, natural and forced vibrations of reinforced shells, and vibrations and acoustics of complex shell structures. He has published more than 150 scientific papers, four monographs, and has six author's certificates and patents.

 **D.V. Shchitov** is the Head of the Department of Construction of the branch of the North-Caucasus Federal University (NCFU) in Pyatigorsk. He is also Associate Professor, candidate of Technical Sciences. He graduated from the State University of Civil Engineering in Rostov-on-Don in 1995. During his training, he participated in research related to mathematical modeling of engineering structures. Since 1997, he worked at Pyatigorsk State Technological University (PSTU) as an Engineer, then as a Teacher at the Faculty of Technology. He received a second higher education in economics in absentia. He studied the postgraduate course of Rostov State University at the Faculty of Mechanics and Mathematics. In 2007, he defended his dissertation on the topic "Statics and vibrations of rotation shells containing liquid". In 2004, he was appointed Acting Head of the Department of Industrial and Civil Engineering. The Department of Construction was established in 2011 on the basis of two adjacent departments. In 2012, the department became part of the Faculty of Engineering of the Institute of Service, Tourism and Design (branch) NCFU in Pyatigorsk. During his tenure as Head of the Department, eight construction laboratories were created and equipped. He is married and has four children. His research interests are in the following areas: technical expertise of buildings, energy audit, construction forensic examination, reconstruction and restoration of buildings, reconstruction of buildings and urban areas, construction, housing, and communal services, real estate valuation, and mathematical and computational models of buildings and structures. He has more than 70 publications, and his Hirsch index is 8.

Contents

Introduction

Mathematical modeling of large shell deformations is relevant in a number of technical fields. First of all, these are the traditional technological challenges of producing a predetermined shape from relatively simple workpieces. Methods of deep drawing, stamping, bending, flanging, running and others are used.

The problems of large plastic shell forming are related to the complex problem of nonlinear mechanics. It is necessary to take into account both geometric and physical nonlinear factors here.

The solution of problems of forming by direct numerical methods is very difficult both in time and in the stability of the account. Although in the case of complex geometry without direct numerical methods such as FEM is indispensable. There are a lot of publications with the use of FEM-packages. But there are few discussions of emerging problems and the effectiveness of their overcoming, for example, by the time of calculations.

Experiments play a significant role in the identification of the defining equations. There are works in which there is a good agreement between the calculations and the performed tests. In works [1, 2], it is noted that in many processes of metal deformation the mechanical characteristics due to high loads become anisotropic. Many engineering applications and calculations require consideration of the strain induced anisotropy of mechanical properties. One of the theories that takes into account such anisotropy is the theory of isotropic translation of the center of the flow surface. The paper [1]

1

describes this theory and indicates its advantages in comparison with other theories of anisotropic plasticity. The experimental method of determination of material parameters is described. A comparison with experimental data, which showed a good match, is given.

A number of widely used materials require constants and ratios characterizing anisotropic hardening. In [3], such information is provided that can be used at the analysis of processes of sheet and volume stamping.

When processing metals, one of the effects is heating. In the monograph [4] the characteristics of metals at elevated temperatures are given; the methods of determining the parameters of the equation of state under hot forming are given. To study the processes of hot forming of metals, the equations of state of rheonomic bodies, which include the structural parameter of damage, are applied. The variants of FEM in the form of displacement method and mixed method, as well as effective analytical methods for solving technological problems of creep theory are proposed.

A modified version of the governing relations of the plasticity theory for initial anisotropic media is proposed and experimentally confirmed in [5]. The presented version of the deformation theory of plasticity of anisotropic media allows us to calculate the ultimate deformation of the material in the technological processes of material processing by pressure.

In [6], localization of plastic deformation is considered as the process of evolution of the active nonlinear medium, accompanied by the generation of a special type of periodic processes — autowaves of localized plastic deformation. The basic differential equations for the kinetics of changes in the autocatalytic (deformation) and damping (stress) factors that control the formation of autowaves in an active nonlinear medium and the main quantitative characteristics of such autowave processes are obtained. The processes of production, rearrangement and destruction of autowaves during plastic deformation are experimentally investigated.

The paper [7] presents the results of experimental verification of the model of anisotropy of ultimate plasticity, describing the destruction of sheet materials with different degrees of anisotropy. A method for determining the material constants included in the model of anisotropy of ultimate plasticity for sheet materials with isotropic plastic deformation resistance is developed.

In the thesis [8], the mathematical elements of the theory of constitutive equations for plasticity at finite deformations are developed. We develop and implement a method that allows us to generalize the defining relations that take place at small deformations to the case of finite deformations based on the use of tensor measures of finite deformations and stresses of various types.

In [9], the general scheme allowing the search for the solution of inverse problems of identification of defining relations by results of technological experiments is proposed. The main feature of this scheme is the development of simplified mathematical models of technological processes of pressure treatment in order to use them in determining the permanent material on the results of test experiments, which are carried out directly on the process equipment. This scheme is implemented on the example of superplastic forming of cylindrical and spherical shells from sheet blanks of titanium alloy.

The coordination of theory and physical experiment is important for the improvement of manufacturing technology and the prediction of critical buckling pressures (actuation pressures) of shells of the type of safety membranes (SM, bursting disks). They belong to the elements of devices of safety systems that protect process equipment and tanks from destruction by overpressure.

The task of creating a high accuracy SM was solved by creating a facility for the manufacture of membranes and the development of equipment for predicting the actuation pressure (AP) by means of non-destructive methods that are currently used. The method of pressure–displacement curve analysis for hardware extrapolation of critical loads is the most convenient for AP prediction. The methods of computer control and mathematical processing of information by software and hardware complex are used [10].

The need to take into account a number of deviations from the spherical ideality of the dome obtained by plastic drawing by uniform pressure from the round clamped plate is indicated in [11]. The following factors are taken into account: the presence of a narrow boundary zone of negative Gaussian curvature, the deviation of the main surface of the dome from the spherical one and thickness variation. On the basis of the approach [11], computational experiments were carried out in [12, 13]. They showed a tendency of convergence of the calculated bifurcation and experimental critical loads of flapping membranes, losing stability in asymmetric forms.

To achieve high quality of SM, the concept and technology of artification has been used recently [14–16]. From the standpoint of the sensitivity of the shells to the initial random technological imperfections, this approach can be interpreted as artificially introduced "imperfections" that overlap the influence of random ones, stabilize the critical load and the form of loss of stability.

In works [17–21], the progress on a way of convergence of the theory and experiment is made. It was possible to build analytical solutions for the problem of forming artificated shells and theoretically substantiate the concept of artification. A mathematical model of deformation of physically nonlinear shells of rotation at large displacements and angles of rotation is presented. The defining relations of Davis–Nadai type are constructed, taking into account the inhomogeneity of the material properties induced by deformations in thickness. Analytical solutions of problems of plastic forming of spherical and ellipsoidal domes from a plate are given. A comparison of theory and experiment was made, which showed their good agreement.

The problems of large deformations of shells of rotation with free drawing pressure are presented in this monograph. The material of shells are considered to be steel-type metals. Therefore, their physical properties are modeled on the basis of nonlinear diagrams of the material, which set the stress intensity as a function of logarithmic deformations. The used mathematical model is based on E. Reissner's equilibrium equations generalized to large elongations. To build physical relations of efforts with the deformations, the Davis–Nadai relations are used.

To solve the problems a semi-inverse method is developed, the technique of which has options depending on the geometry of the shells. The applications of the method to the molding of high-precision membrane devices used to protect structures from destruction by overpressure are shown; to optimize the design of containers for the transport of liquid cargo; to the modeling of the work of flat jacks used for lifting and leveling heavy buildings and structures.

In Chapter 1, we derive equations for solving problems of large changes in the initial shapes of rotation shells with large relative elongations (deformations). The change in the metric of the middle surface and the transverse compression of the shell in thickness are taken into account. It is based on the equations of E. Reissner, in which large displacements and rotation angles are allowed, but the deformations are assumed to be small.

The Chapter 2 is devoted to an important aspect of the problems of stretching under large deformations — the construction of physical relationships between internal forces and relative elongations. The plastic properties of a material are determined by a loading diagram that links the intensity of stress and the intensity of logarithmic deformations. The concept of a secant module is used. The defining relations closing the system of resolving equations are derived.

Chapter 3 introduces the flapping membrane type of shells. They are convex domed shells used to protect equipment from destruction by excessive internal pressure. The information related to modeling tasks and manufacturing technologies for high-precision flapping safety membranes is presented. Analysis of experiments in this area has revealed the necessary information for the use of the semi-inverse method.

In Chapter 4, the semi-inverse method is tested on the problems of forming a spherical dome and an ellipsoid shell with a small Meridian eccentricity. A comparison of theoretical and experimental results obtained by applying real molding technology is given.

In Chapter 5, the semi-inverse method is applied to the problem of drawing an ellipsoidal dome without restrictions on the smallness of eccentricity. The application of the method to the problem of inflating a cylindrical shell into a barrel-shaped one is also shown.

Chapter 6 introduces tanks for transporting dangerous liquid cargo. The problems of strength leading to the destruction of tank bottoms during transportation are shown. In the version of the tank used for calculations, the bottom of the composite geometry is parameterized. The stress state of the bottom loaded with the weight of the liquid is studied. Calculations and analysis of static stress states of bottoms of different geometries are performed. We consider shells with a smoothly changing Meridian, with a Meridian line that has fractures, and with an annular groove. A comparison of the theory and experiment is given, and the appearance of differently oriented cracks is explained. A semi-inverse method was used to change the original shape of the tank bottom to a more optimal shape.

Chapter 7 discusses the specifics of using jacking systems in construction. In particular, in technologies of lifting and leveling buildings that have received a roll. A comparison of piston and flat jacks is given. Possible variants of meridian forms of flat jack shells are considered and the stress state at the initial loading stage is analyzed. Next, the simulation is performed using the semi-inverse method of flat jack operation at large movements.

In the Chapter 8, the equations of statics of axisymmetric deformation of the shells of rotation are presented. Two variants of geometrically nonlinear quadratic approximation equations are considered: in the coordinate system of the accompanying trihedron and in the cylindrical coordinate system. The interface conditions on ring edges are written. A variant of the theory of shells of the E. Reissner type with large rotation angles is also given. To apply algorithms for solving boundary value problems, the equations are reduced to the canonical form. An example of solving boundary value problems in the theory of shells in the calculation of sealing elements of hydraulic connections is considered.

In Conclusion, we pay homage to integrated packages that combine the capabilities of symbolic and computational mathematics, programming tools, graphics and animation. The MathCad series packages that are used to implement the methods in this book have a particularly user-friendly interface. It allows you to create programs that are easy to understand, modify, and pass on to other users. This opens up new opportunities for implementing programs for educational, scientific and research purposes, and creating appropriate libraries.

Chapter 1

Generalization of E. Reissner's Equations for the Problem of Shells of Rotation Deformation

E. Reissner's equations [41–44] are a suitable tool for the analysis of axisymmetric deformation of shells at large displacements and angles of rotation, since they do not impose restrictions on these kinematic characteristics. Despite the fact that they are derived for small deformations, they can be generalized to the case of large elongations, leading to a change in the metric of the coordinate surface of the deformed shell and compression of the normal. This chapter presents the corresponding kinematic relations. On the basis of the direct normal hypothesis, taking into account the transverse shift, expressions of elementary work of internal forces are obtained. The equilibrium equations of the forces and moments are derived as a consequence of the principle of the possible moving of Lagrange.

1.1. Surface Geometry of Rotation and Spatial Curvilinear Coordinate System

In a rectangular Cartesian coordinate system X, Y, Z, we consider the surface of rotation O_o, for which the Z axis is the axis of symmetry. Surface O_o is further considered as the middle surface of a shell, the coordinates of the points of which are determined by the radius vector

$$r^o(\xi, \theta) = r_o(\xi)i(\theta) + z_o(\xi)i_3, \qquad (1.1.1)$$

where

$$i = i_1 \cos\theta + i_2 \sin\theta, \qquad (1.1.2)$$

i_k are the unit vectors of the axes of a Cartesian system ($k = 1, 2, 3$) and $\xi = q^1 = \alpha_1$, $\theta = q^2 = \alpha_2$ are orthogonal curvilinear (Gaussian) coordinates on the surface of O_o, the grid of which coincides with the main curvature lines.

The basic trihedron and the coefficients of the first quadratic form of the surface are determined as follows [33]:

$$r^o_s = \partial r^o/\partial q^s, \quad (s, k = 1, 2).$$

$$r^o_1 = z'_o i_3 + r'_o i, \quad r^o_2 = r_o(-i_1 \sin\theta + i_2 \cos\theta);$$

$$g^o_{11} = (r'_o)^2 + (z'_o)^2 = (\alpha_o)^2, \quad g^o_{12} = 0, \quad g^o_{22} = (r_o)^2;$$

$$(\cdots)' = d(\cdots)/d\xi = (\cdots)_{,\xi}, \quad \xi = q_1. \qquad (1.1.3)$$

Due to orthogonality of basis vectors r^o_s component $g^o_{12} = 0$, the coefficients of the first quadratic form correspond to the components of the first metric surface tensor. Values $\alpha_o = (g^o_{11})^{1/2} = |r^o_1|$ $\alpha_o = (g^o_{22})^{1/2} = |r^o_2|$ are the Lame coefficients of coordinate lines.

We introduce the unit vectors of coordinate lines and the unit vector of the normal to the surface O_o

$$e^o_1 = r^o_1/\alpha_o, \quad e^o_2 = r^o_2 r_o \qquad (1.1.4)$$

and the unit vector of the normal $n_o = e^o_3$ to the surface O_o, which is determined by the vector product

$$e^o_3 = e^o_1 \times e^o_2 = (-z'_o i + r'_o i_3)/\alpha_o. \qquad (1.1.5)$$

We denote $\Phi_o = \langle i, e^o_1 \rangle$, the angle between the vectors i and e^o_1, or between the internal normal to the surface and the axis of symmetry Z, Fig. 1.1.1.

The following relations exist:

$$\cos\Phi_o = r'_o/\alpha_o, \quad \sin\Phi_o = z'_o/\alpha_o. \qquad (1.1.6)$$

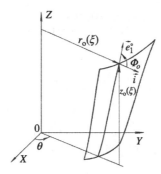

Fig. 1.1.1. Coordinate system.

The coefficients of the second quadratic form are determined by scalar products

$$b_{sk}^o = e_3^o \cdot \frac{\partial^2 r^o}{\partial q^s \partial q^k}, \tag{1.1.7}$$

and, taking into account (1.1.6), they have the form

$$b_{11}^o = \alpha_o \Phi_o', \quad b_{12}^o = 0, \quad b_{22}^o = r_o \sin \Phi_o. \tag{1.1.8}$$

The main curvatures of the surface are the roots of the characteristic equation

$$det[\|b_{sk}^o\| - k^o \|g_{sk}^o\|] = 0, \tag{1.1.9}$$

from which

$$k_1^o = \Phi_o' \alpha_o, \quad k_2^o = (\sin \Phi_o) r_o. \tag{1.1.10}$$

We have the derivation formula for an arbitrary surface related to the lines of the main curvatures

$$\frac{\partial e_1^o}{\partial q^1} = -\frac{1}{H_2^o} \frac{\partial H_1^o}{\partial q^2} e_2^o + H_1^o k_1^o e_3^o, \quad \frac{\partial e_1^o}{\partial q^2} = \frac{1}{H_1^o} \frac{\partial H_2^o}{\partial q^1} e_2^o,$$

$$\frac{\partial e_3^o}{\partial q^1} = -H_1^o k_1^o e_1^o, \quad \frac{\partial e_3^o}{\partial q^2} = -H_2^o k_2^o e_2^o,$$

$$\frac{\partial e_2^o}{\partial q^1} = \frac{1}{H_2^o} \frac{\partial H_1^o}{\partial q^2} e_1^o, \quad \frac{\partial e_2^o}{\partial q^2} = -\frac{1}{H_1^o} \frac{\partial H_2^o}{\partial q^1} e_1^o + H_2^o k_2^o e_3^o, \tag{1.1.11}$$

where H_1^o, H_2^o are Lamé coefficients, $q^1 = \xi, q^2 = \theta$. Further, the partial derivatives are denoted more compactly, similarly (1.1.11):

$$(\cdots)_{,\xi} = \partial(\cdots)/\partial\xi, \quad (\cdots)_{,\theta} = \partial(\cdots)/\partial\theta. \tag{1.1.12}$$

Taking into account the introduced designations on the surface O_o, the derivational formulas have the form

$$e^o_{1,\xi} = \alpha_o k^o_1 e^o_3, \quad e^o_{1,\theta} = \cos \Phi_o e^o_2,$$

$$e^o_{2,\xi} = 0, \quad e^o_{2,\theta} = - \cos \Phi_o e^o_1 + r_o k^o_2 e^o_3,$$

$$e^o_{3,\xi} = -\alpha_o k^o_1 e^o_1, \quad e^o_{3,\theta} = -r_o k^o_2 e^o_2. \tag{1.1.13}$$

We introduce a special spatial orthogonal system of curvilinear coordinates $\xi, \theta, \zeta = q^3$, the points of space which will be defined by a radius-vector

$$\boldsymbol{R}^o = \boldsymbol{r}^o + \zeta e^o_3. \tag{1.1.14}$$

Using derivational formulas (1.1.13), we obtain

$$\boldsymbol{R}^o_1 = \alpha^\zeta_o e^o_1, \quad \boldsymbol{R}^o_2 = r^\zeta_o e^o_2, \quad \boldsymbol{R}^o_3 = e^o_3; \tag{1.1.15}$$

$$G^o_{11} = (\alpha^\zeta_o)^2, \quad G^o_{22} = (r^\zeta_o)^2, \quad G^o_{33} = 1;$$

$$G^o_{sk} = 0, \quad s \neq k,$$

$$G^o = det \, \|G^o_{sk}\| = (\alpha^\zeta_o)^2 (r^\zeta_o)^2, \tag{1.1.16}$$

where

$$\alpha^\zeta_o = \alpha_o(1 - \zeta k^o_1), \quad r^\zeta_o = r_o(1 - \zeta k^o_2). \tag{1.1.17}$$

Here, $\boldsymbol{R}^o_s$ is the main basis and G^o_{sk} are covariant components of the metric tensor $\hat{G}^o$ of the reference system (1.1.15).

1.2. Kinematics of Axisymmetric Deformation

Let O_o be the middle surface of the thin shell that fills the volume V_o before deformation. Let's consider an axisymmetric deformation in which the displacement vector V of each point of the shell lies in the meridional section.

We introduce the following hypotheses:

(a) the material normal during the process of deformation remains rectilinear, but non-orthogonal to the middle surface of the deformed shell (the transverse shift is taken into account).

(b) the compression of the material normal is considered uniform over the thickness of the shell, i.e. it is independent of the transverse coordinate ζ, so that $\varepsilon^\zeta_3 \approx \varepsilon_3$ on the middle surface.

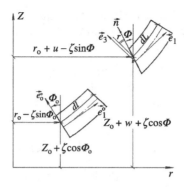

Fig. 1.2.1.   The deformation of the elementary volume of the shell.

(c) the angle of transverse shear γ^ζ between the directions of the
lines ξ and ζ is small to be able to put $\cos\gamma^\zeta \approx 1, \sin\gamma^\zeta \approx \gamma^\zeta$.

The transverse shear is taken into account as the amount of dis-
tortion of the right angle between the normal n and the tangent
planes of equidistant surfaces, Fig. 1.2.1.

The introduction of hypotheses is equivalent to the imposition of
certain geometric relationships on the shell material. In comparison
with the hypotheses of the Kirchhoff–Love, the connections between
hypotheses (a) and (c) are less stringent because the possibility of
deflection of the material of the normal from the geometric introduces
an additional degree of freedom. Therefore, without complicating the
basic equations, taking into account the transverse shear in this form
gives certain advantages.

First, the known contradiction of the theory of shells based on
Kirchhoff–Love hypotheses is eliminated, when the effect of tangential stresses is taken into account, and the shift is considered to be
zero.

Second, the derivation of equilibrium equations is simplified from
the principle of possible displacements, as the number of independent
variations increases.

After the deformation of the surface, O_o moves to some surface of
rotation O. Since the material normal to the original surface may not
remain orthogonal to the middle surface after deformation, the angle
should be distinguished $\Phi = \langle i_3, n \rangle$ between the axis of symmetry
of the shell and the material normal n, and angle, $\Phi + \gamma = \langle i, e_1 \rangle$

between the polar radius and the vector tangent to the surface meridian O. Here, $\gamma = \gamma^\varsigma|_{\varsigma=0}$ — is the transverse shear angle on the middle surface.

In the designations of the main geometric characteristics of the deformed shell, we use the same analogs as above, but without zeros. After deformation, the volume V_o goes to V, radius-vector r^o goes to r

$$r(\xi,\theta) = r(\xi)i(\theta) + z(\xi)i_3. \tag{1.2.1}$$

For surface O, there are type formulas presented in Section 1.1, as follows:

$$r_s = \partial r/\partial q^s, \quad (s,k) = 1,2.$$

$$r_1 = z'i_3 + r'i, \quad r_2 = r(-i_1 \sin\theta + i_2 \cos\theta);$$

$$g_{11} = (r')^2 + (z')^2 = \alpha^2, \quad g_{12} = 0, \quad g_{22} = r^2;$$

$$e_1 = r_1/\alpha, \quad e_2 = r_2/r, \quad e_3 = e_1 \times e_2 = (-z'i + r'i_3)/\alpha,$$

$$\cos(\Phi + \gamma) = r'/\alpha, \quad \sin(\Phi + \gamma) = z'/\alpha,$$

$$b_{11} = \alpha(\Phi + \gamma)', \quad b_{12} = 0, \quad b_{22} = r \cdot \sin(\Phi + \gamma),$$

$$k_1 = (\Phi + \gamma)'\alpha, \quad k_2 = \sin(\Phi + \gamma)/r. \tag{1.2.2}$$

Differentiation formulas have the form

$$e_{1,\xi} = \alpha k_1 e_3, \quad e_{1,\theta} = \cos(\Phi + \gamma)e_2,$$

$$e_{2,\xi} = 0, \quad e_{2,\theta} = -\cos(\Phi + \gamma)e_1 + rk_2 e_3,$$

$$e_{3,\xi} = -\alpha k_1 e_1, \quad e_{3,\theta} = -rk_2 e_2. \tag{1.2.3}$$

In the process of deformation the radius vector R^o becomes a vector R

$$R = r + \zeta(1 + \varepsilon_3)n = r + \zeta(1 + \varepsilon_3)(e_1 \sin\gamma + e_3 \cos\gamma). \tag{1.2.4}$$

The vectors of the Lagrangian basis of the deformed shell, taking into account the small angle of shift γ, have the form

$$R_1 = \alpha^\varsigma e_1 + \zeta(1 + \varepsilon_3)\alpha\gamma K_1 e_3; \quad R_2 = r^\varsigma e_2; \quad R_3 = (1 + \varepsilon_3)\gamma e_1,$$

$$G_{11} = (\alpha_o^\varsigma)^2(1 + \varepsilon_1^\varsigma)^2, \quad G_{22} = (r_o^\varsigma)^2(1 + \varepsilon_2^\varsigma)^2,$$

$$G_{33} = (1 + \varepsilon_3)^2, \quad G_{13} = \alpha_o^\varsigma \gamma(1 + \varepsilon_1^\varsigma)(1 + \varepsilon_3), \tag{1.2.5}$$

where

$$\alpha^{\zeta} = \alpha(1 - \zeta\bar{k}_1), \quad r^{\zeta} = r(1 - \zeta\bar{k}_2),$$

$$\bar{k}_1 = \Phi'/\alpha, \quad \bar{k}_2 = (\sin\Phi)/r. \tag{1.2.6}$$

Through the scalar products of the vectors of the main Lagrangian basis of the deformed shell, it is rather difficult to derive expressions of the components of the first strain measure with a clear geometric meaning. The approach used by E. Reissner to derive equations with large rotation angles and small deformations is more convenient [42, 43]. This way allows you to perform a generalization of kinematics and equilibrium equations to account for large deformations, taking into account normal compression.

We derive expressions for the components of relative elongations. According to Fig. 1.2.1, for the circumferential direction we have

$$\varepsilon_2^{\zeta} = \frac{\{2\pi[r_o + u - \zeta(1 + \varepsilon_3)\sin\Phi] - 2\pi(r_o - \zeta\sin\Phi_o)]\}}{[2\pi(r_o - \zeta\sin\Phi_o)]}, \tag{1.2.7}$$

or

$$\varepsilon_2^{\zeta} = (\varepsilon_2 + \zeta\kappa_2)(1 - \zeta k_2^o), \quad \varepsilon_2 = u/r_o,$$

$$\kappa_2 = k_2^o - (1 + \varepsilon_3)K_2, \quad k_2^o = (\sin\Phi_o)/r_o,$$

$$K_2 = (\sin\Phi)/r_o. \tag{1.2.8}$$

To derive the components of the meridional elongation and the angle of the transverse shear, we use the differential relations, which explains Fig. 1.2.1 in vertical and horizontal axis projections,

$$(1 + \varepsilon_1^{\zeta})\cos(\Phi + \gamma)dl = d[r_o + u - \zeta(1 + \varepsilon_3)\sin\Phi]_1,$$

$$(1 + \varepsilon_1^{\zeta})\sin(\Phi + \gamma)dl = d[r_o + w + \zeta(1 + \varepsilon_3)\cos\Phi]_2, \tag{1.2.9}$$

where

$$dl = \alpha_o(1 - \zeta k_1^o)d\xi, \quad d(\cdots) = (\cdots)'d\xi.$$

Let's open differentials of square brackets

$$d[\cdots]_1 = [\cdots]_1'd\xi$$

$$= [r_o' + u' - \zeta(1 + \varepsilon_3)\Phi'\cos\Phi - \zeta\varepsilon_3'\sin\Phi]_3 d\xi,$$

$$d[\cdots]_2 = [\cdots]_2'd\xi$$

$$= [r_o' + w' - \zeta(1 + \varepsilon_3)\Phi'\sin\Phi + \zeta\varepsilon_3'\cos\Phi]_4 d\xi. \tag{1.2.10}$$

From the relations (1.2.10), we obtain two other equalities by performing the following operations. At the beginning, we multiply the first ratio by $\cos\Phi$, second by $\sin\Phi$, and sum. Then we multiply the first by $\sin\Phi$, second by $\cos\Phi$, and subtract the first from the second. After performing the transformation taking into account the smallness γ, we obtain

$$\alpha_o(1 - \zeta k_1^o)(1 + \varepsilon_1^\zeta)$$
$$= \{[\cdots]_3 \cos\Phi + [\cdots]_4 \sin\Phi\} = \alpha_o(1 + \varepsilon_1) - \zeta(1 + \varepsilon_3)\Phi',$$
$$\alpha_o(1 - \zeta k_1^o)(1 + \varepsilon_1^\zeta)\gamma$$
$$= \{[\cdots]_4 \cos\Phi - [\cdots]_3 \sin\Phi\} = \alpha_o\gamma_o + \zeta\varepsilon_3', \qquad (1.2.11)$$

where

$$\alpha_o(1 + \varepsilon_1) = w' \sin\Phi + u' \cos\Phi + \alpha_o \cos(\Phi - \Phi_o),$$
$$\alpha_o\gamma_o = w' \cos\Phi - u' \sin\Phi - \alpha_o \sin(\Phi - \Phi_o). \qquad (1.2.12)$$

Thus,

$$\alpha_o(1 - \zeta k_1^o)(1 + \varepsilon_1^\zeta) = \alpha_o(1 + \varepsilon_1) - \zeta(1 + \varepsilon_3)\Phi',$$
$$\alpha_o(1 - \zeta k_1^o)(1 + \varepsilon_1^\zeta)\gamma = \alpha_o\gamma_o + \zeta\varepsilon_3'. \qquad (1.2.13)$$

From ratios (1.2.12), (1.2.13), it follows that

$$\varepsilon_1^\zeta = (\varepsilon_1 + \zeta\kappa_1)/(1 - \zeta k_1^o),$$
$$\varepsilon_1 = (w' \sin\Phi + u' \cos\Phi)/\alpha_o + \cos(\Phi - \Phi_o) - 1$$
$$\kappa_1 = k_1^o - (1 + \varepsilon_3)K_1, \quad k_1^o = \Phi_o'/\alpha_o, \quad K_1 = \Phi'/\alpha_o,$$
$$\gamma = (\alpha_o\gamma_o + \zeta\varepsilon_3')/[\alpha_o(1 + \varepsilon_1) - \zeta(1 + \varepsilon_3)\Phi']. \qquad (1.2.14)$$

The denominator in the right part of the expression for γ in (1.2.14) is represented as

$$[\ldots] = \alpha_o(1 + \varepsilon_1)[1 - \zeta(1 + \varepsilon_3)\bar{k}_1],$$

where

$$\bar{k}_1 = \Phi'/[\alpha_o(1 + \varepsilon_1)] = \Phi'/\alpha$$

curvature of the meridian of the shell is in the deformed state. We consider such a deformation during which $-1 < \varepsilon_3 \ll 1$, and the original and deformed shell are thin, so $\zeta k_1^o \ll 1, \zeta(1 + \varepsilon_3)\bar{k}_1 \ll 1$. Therefore, (1.2.14) can be simplified as

$$\varepsilon_1^\zeta = (\varepsilon_1 + \zeta\kappa_1), \quad \gamma = (\gamma_o + \zeta\varepsilon_3'/\alpha_o)/(1 + \varepsilon_1). \tag{1.2.15}$$

Thus, we have the following summary of formulas for kinematic relations:

$$\varepsilon_1^\zeta = (\varepsilon_1 + \zeta\kappa_1), \quad \varepsilon_2^\zeta = (\varepsilon_2 + \zeta\kappa_2), \quad \varepsilon_2 = u/r_o,$$

$$\varepsilon_1 = (w'\sin\Phi + u'\cos\Phi)/\alpha_o + \cos(\Phi - \Phi_o) - 1,$$

$$\kappa_1 = k_1^o - (1 + \varepsilon_3)K_1, \quad k_1^o = \Phi_o'/\alpha_o, K_1 = \Phi'/\alpha_o,$$

$$K_2 = (\sin\Phi)/r_o,$$

$$\kappa_2 = \kappa_2^o - (1 + \varepsilon_3)K_2, \quad k_2^o = (\sin\Phi_o)/r_o,$$

$$\gamma = \gamma_o/(1 + \varepsilon_1) + \zeta\varepsilon_3'/[\alpha_o(1 + \varepsilon_1)],$$

$$\gamma_o = (w'\cos\Phi - u'\sin\Phi)/\alpha_o - \sin(\Phi - \Phi_o). \tag{1.2.16}$$

It also follows from the ratios (1.2.16) that

$$u' \equiv (r_o\varepsilon_2)' = \alpha_o(\varepsilon_1\cos\Phi - \gamma_o\sin\Phi + \cos\Phi - \cos\Phi_o),$$

$$w' = \alpha_o(\varepsilon_1\sin\Phi + \gamma_o\cos\Phi + \sin\Phi - \sin\Phi_o). \tag{1.2.17}$$

The first of the relations (1.2.17) is a condition of compatibility of deformations.

The components of the first measure of deformation are related to the elongation formulas

$$G_{11} = (\alpha_o^\zeta)^2(1 + \varepsilon_1^\zeta)^2, \quad G_{22} = (r_o^\zeta)^2(1 + \varepsilon_2^\zeta)^2,$$

$$G_{33} = (1 + \varepsilon_3)^2, \quad G_{13} = \alpha_o^\zeta\gamma(1 + \varepsilon_1^\zeta)(1 + \varepsilon_3);$$

$$G = det\|G_{sk}\|$$

$$= (\alpha_o^\zeta)^2(r_o^\zeta)^2(1 + \varepsilon_1^\zeta)^2(1 + \varepsilon_2^\zeta)^2(1 + \varepsilon_3)^2. \tag{1.2.18}$$

where

$$\alpha_o^\zeta = \alpha_o(1 - \zeta k_1^o), \quad r_o^\zeta = r_o(1 - \zeta k_2^o). \tag{1.2.19}$$

The elementary volume of the deformed state is

$$dV = \sqrt{G}d\zeta d\xi d\theta = \alpha_o^\zeta r_o^\zeta(1 + \varepsilon_1^\zeta)(1 + \varepsilon_2^\zeta)(1 + \varepsilon_3)d\zeta d\xi d\theta. \quad (1.2.20)$$

The elementary volume in the undeformed state is

$$dV_o = \sqrt{G^o}d\zeta d\xi d\theta = \alpha_o^\zeta r_o^\zeta d\zeta d\xi d\theta. \quad (1.2.21)$$

1.3. Virtual Work of the Internal Forces

For the transition to the integral characteristics of the stress state, we consider the elementary work of internal forces on virtual displacements, which in the original spatial version at large strains has the form [33]

$$\widehat{\delta}a_{(i)} = -0.5 \iiint_V t^{sk}(\delta G_{sk})dV, \quad (1.3.1)$$

where t^{ij} are contravariant components of the Cauchy stress tensor

$$\hat{T} = t^{sk}\boldsymbol{R}_s\boldsymbol{R}_k. \quad (1.3.2)$$

In the considered variant of strain,

$$\tilde{\delta}a_{(i)} = -\iiint_V [0.5t^{11}(\delta G_{11}) + 0.5t^{22}(\delta G_{22})$$
$$+ 0.5t^{33}(\delta G_{33}) + t^{13}(\delta G_{13})]dV. \quad (1.3.3)$$

Let's move on to the tensor representation $\hat{T}$ in the form

$$\hat{T} = \sum_{s,k=1}^{3} t_{(sk)}\boldsymbol{E}_s\boldsymbol{E}_k, \quad (1.3.4)$$

where $t_{(sk)} = t^{sk}(G_{ss}G_{kk})^{1/2}$ (do not sum by s and k) and $\boldsymbol{E}_k = \boldsymbol{R}_k/(G_{kk})^{1/2}$ are unit vectors of the Lagrangian basis of the deformed state.

In the absence of the shift, the vectors $\boldsymbol{E}_k$ constitute an orthogonal basis, and the values of $t_{(sk)}$ are the physical components of the stress tensor $\hat{T}$. In the presence of a transverse shear, the basis $\boldsymbol{E}_k$ is oblique. In this case, you can determine that $t_{(sk)}$ are the components of the decomposition in the basis $\boldsymbol{E}_k$ of the vectors of voltages $\overset{s}{\boldsymbol{t}}$ at sites which had, prior to deformation, normal isolated $\boldsymbol{E}_s^o$. Indeed, according to [33] for elementary oriented site $\boldsymbol{E}_1^o \, d\overset{1}{O}{}^o = (\boldsymbol{R}_1^o/\alpha_o^\varsigma)$, we have in the deformed state

$$\overset{1}{t} \, d\overset{1}{O} = \sqrt{G/G^o}\, t^{sk} \boldsymbol{R}_k \boldsymbol{R}_s^o \bullet \boldsymbol{R}_1^o d\overset{1}{O}_o /\alpha_o^\varsigma$$

$$= \sqrt{G/G^o}\, \alpha_o^\varsigma (t^{11} \boldsymbol{R}_1 + t^{13} \boldsymbol{R}_3) d\overset{1}{O}_o, \qquad (1.3.5)$$

$$d\overset{1}{O} = r_o^\varsigma (1 + \varepsilon_2^\varsigma)(1 + \varepsilon_3) d\theta d\varsigma, \quad d\overset{1}{O}_o = r_o^\varsigma d\theta d\varsigma,$$

$$\sqrt{G/G^o} = (1 + \varepsilon_2^\varsigma)(1 + \varepsilon_2^\varsigma)(1 + \varepsilon_3), \quad \overset{1}{t} = t_{(11)} \boldsymbol{E}_1 + t_{(13)} \boldsymbol{E}_3. \qquad (1.3.6)$$

Analogically,

$$d\overset{2}{O} = \alpha_o^\varsigma (1 + \varepsilon_1^\varsigma)(1 + \varepsilon_3) d\xi d\varsigma, \quad d\overset{2}{O}_o = r_o^\varsigma d\xi d\theta,$$

$$\overset{2}{t} = t_{(22)} \boldsymbol{E}_2. \qquad (1.3.7)$$

Let us reveal the terms of integrands (1.3.3)

$$0.5 t^{11}(\delta G_{11}) dV = (1 + \varepsilon_3) t_{(11)} \Delta_2 [\delta \varepsilon_1 + \varsigma(1 + \varepsilon_3) \delta \kappa_1$$

$$- \varsigma K_1 \delta \varepsilon_3 + \gamma (\delta \gamma_o + \varsigma \delta \gamma_1)] \alpha_o r_o d\varsigma d\xi d\theta,$$

$$0.5 t^{22}(\delta G_{22}) dV = (1 + \varepsilon_3) t_{(22)} \Delta_1 [\delta \varepsilon_2 + \varsigma(1 + \varepsilon_3) \delta \kappa_2$$

$$- \varsigma K_2 \delta \varepsilon_3 + \gamma (\delta \gamma_o + \varsigma \delta \gamma_1)] \alpha_o r_o d\varsigma d\xi d\theta,$$

$$0.5 t^{33}(\delta G_{33}) dV = t_{(33)} \Delta_1 \Delta_2 (\delta \varepsilon_3) \alpha_o r_o d\varsigma d\xi d\theta,$$

$$t^{13}(\delta G_{13}) dV = [(1 + \varepsilon_3) t_{(13)} \Delta_2 (\delta \gamma_o + \varsigma \delta \gamma_1) + t_{(13)} \Delta_1 \Delta_2$$

$$+ \alpha_o \gamma^\varsigma (\delta \varepsilon_3)] \alpha_o r_o d\varsigma d\xi d\theta. \qquad (1.3.8)$$

The symbols are as follows:

$$\Delta_1 = (1 + \varepsilon_1^\varsigma)(1 - \varsigma k_1^o), \quad \Delta_2 = (1 + \varepsilon_2^\varsigma)(1 - \varsigma k_2^o),$$

$$K_1 = \Phi'/\alpha_o, \quad K_2 = (\sin \Phi)/r_o, \quad \gamma_1 = \varepsilon_3'/\alpha_o. \qquad (1.3.9)$$

The shells are believed to be thin in the initial state of, so that $\varsigma k_1^o \ll 1$, $\varsigma k_2^o \ll 1$.

So we can put

$$\Delta_1 = (1 + \varepsilon_1^\zeta), \quad \Delta_2 = (1 + \varepsilon_2^\zeta). \tag{1.3.10}$$

By collecting the coefficients of the variations of the deformation components, we obtain

$$\begin{aligned}
\widehat{\delta a}_{(i)} = -\iint_{O_o} (1 + \varepsilon_3) \int_{-h_o/2}^{h_o/2} \Big\{ & t_{(11)}\Delta_2\delta\varepsilon_1 \\
& + t_{(22)}\Delta_1\delta\varepsilon_2 + (t_{(13)} + t_{(11)}\gamma)\Delta_2\delta\gamma_o \\
& + t_{(11)}\Delta_2(1 + \varepsilon_3)\zeta\delta\kappa_1 + t_{(22)}\Delta_1(1 + \varepsilon_3)\zeta\delta\kappa_2 \\
& - t_{(11)}\Delta_2\zeta K_1\delta\varepsilon_3 - t_{(22)}\Delta_1\zeta K_2\delta\varepsilon_3 \\
& + (1 + \varepsilon_3)^{-1}(t_{(33)} + t_{(13)}\gamma)\Delta_1\Delta_2\alpha_o\delta\varepsilon_3 \\
& + (t_{(13)} + t_{(11)}\gamma)\Delta_2\zeta\delta\gamma_1 \Big\}\alpha_o r_o d\zeta d\xi d\theta. \tag{1.3.11}
\end{aligned}$$

We introduce the integral characteristics of the stress state

$$N_1^o = (1 + \varepsilon_3)\int_{-h_o/2}^{h_o/2} t_{(11)}\Delta_2 d\zeta, \quad N_2^o = (1 + \varepsilon_3)\int_{-h_o/2}^{h_o/2} t_{(22)}\Delta_1 d\zeta,$$

$$M_1^o = (1 + \varepsilon_3)^2\int_{-h_o/2}^{h_o/2} t_{(11)}\Delta_2\zeta d\zeta,$$

$$M_2^o = (1 + \varepsilon_3)^2\int_{-h_o/2}^{h_o/2} t_{(22)}\Delta_1\zeta d\zeta,$$

$$Q^o = (1 + \varepsilon_3)\int_{-h_o/2}^{h_o/2} (t_{(13)} + \gamma t_{(11)})\Delta_2 d\zeta. \tag{1.3.12}$$

In accordance with the hypothesis of Kirchhoff type, the stresses $t_{(33)} + t_{(13)}\gamma$ can be neglected compared to others. Then,

$$\begin{aligned}
\widehat{\delta a}_{(i)} = -\iint_{O_o} \Big\{ & N_1^o\delta\varepsilon_1 + N_2^o\delta\varepsilon_2 + Q^o\delta\gamma_o + M_1^o\delta\kappa_1 \\
& + M_2^o\delta\kappa_2 + (1 + \varepsilon_3)^{-1}[M^o\delta\gamma_1 - (K_1 M_1^o \\
& + K_2 M_2^o)\delta\varepsilon_3]\Big\}\alpha_o r_o d\xi d\theta. \tag{1.3.13}
\end{aligned}$$

To determine ε_3 and exclude the variation $\delta\varepsilon_3$ from (1.3.13), it is necessary to involve additional conditions. Such conditions may be the hypothesis of incompressibility of the material or the condition of low normal stresses on equidistant surfaces ($t_{(33)} \approx 0$) in the determining relations.

Under the assumption of incompressibility of the material at the points of the entire shell, including the middle surface, volume conservation relations are performed

$$(1+\varepsilon_1^\varsigma)(1+\varepsilon_2^\varsigma)(1+\varepsilon_3) = 1, \quad (1+\varepsilon_1)(1+\varepsilon_2)(1+\varepsilon_3) = 1, \quad (1.3.14)$$

or

$$\bar{e}_1 + \bar{e}_2 + \bar{e}_3 = 0, \quad \bar{\varepsilon}_1 + \bar{\varepsilon}_2 + \bar{\varepsilon}_3 = 0, \quad (1.3.15)$$

where

$$\bar{e}_1 = \ln(1 + \varepsilon_1^\varsigma), \quad \bar{e}_2 = \ln(1 + \varepsilon_2^\varsigma),$$
$$\bar{\varepsilon}_1 = \ln(1 + \varepsilon_1), \quad \bar{\varepsilon}_2 = \ln(1 + \varepsilon_2), \quad \bar{\varepsilon}_3 = \ln(1 + \varepsilon_3) \quad (1.3.16)$$

— logarithmic (true) deformation.
From (1.3.15), it follows

$$(1 + \varepsilon_3) = (1 + \varepsilon_1)^{-1}(1 + \varepsilon_2)^{-1},$$
$$-(1 + \varepsilon_3)^{-1}\delta\varepsilon_3 = (1 + \varepsilon_1)^{-1}\delta\varepsilon_1 + (1 + \varepsilon_2)^{-1}\delta\varepsilon_2. \quad (1.3.17)$$

After substituting (1.3.17) into (1.3.13) and regrouping the coefficients for the remaining independent variations, we obtain

$$\widehat{\delta a_{(i)}} = -\iint_{O_o} \{\bar{N}_1^o \delta\varepsilon_1 + \bar{N}_2^o \delta\varepsilon_2 + Q^o \delta\gamma_o$$
$$+ M_1^o \delta\kappa_1 + M_2^o \delta\kappa_2\}\alpha_o r_o d\xi d\theta, \quad (1.3.18)$$

where

$$\bar{N}_1^o = N_1^o + (1 + \varepsilon_1)^{-1}(K_1 M_1^o + K_2 M_2^o), \quad (1.3.19)$$
$$\bar{N}_2^o = N_2^o + (1 + \varepsilon_2)^{-1}(K_1 M_1^o + K_2 M_2^o). \quad (1.3.20)$$

In the other version, discussed in Chapter 2, the Kirchhoff hypothesis ($t_{(33)} \approx 0$) and the defining material relations are used to exclude variation $\delta \varepsilon_3$.

1.4. Equilibrium Equations of the Efforts and Moments

To derive the equilibrium equations, we use the principle of possible displacements. In accordance with it, the elementary work of all external and internal forces on the virtual movement of points of the mechanical system from its equilibrium state is zero.

$$\tilde{\delta} a_{(i)} + \tilde{\delta} a_{(e)} = 0, \qquad (1.4.1)$$

where

$$\tilde{\delta} a_{(i)} = -\iint_{O_o} \{ \bar{N}_1^o \delta \varepsilon_1 + \bar{N}_2^o \delta \varepsilon_2 + Q^o \delta \gamma_o $$

$$+ M_1^o \delta \kappa_1 + M_2^o \delta \kappa_2 \} \alpha_o r_o d\xi d\theta,$$

$$\tilde{\delta} a_{(e)} = \iint_{O_o} [p_u (1 + \varepsilon_1)(1 + \varepsilon_2)(\delta u)$$

$$+ p_w (1 + \varepsilon_1)(1 + \varepsilon_2)(\delta w)] \alpha_o r_o d\xi d\theta \qquad (1.4.2)$$

is the work of external forces on virtual movements.

The variations of the deformation components included in (1.4.2), according to their expressions through generalized displacements (1.2.16), have the form

$$\delta \varepsilon_1 = (\delta w)'(\sin \Phi)/\alpha_o + (\delta u)' \cos \Phi / \alpha_o + \gamma_o \delta \Phi,$$

$$\delta \varepsilon_2 = \delta u / r_o,$$

$$\delta \gamma_o = (\delta w)'(\cos \Phi)/\alpha_o - (\delta u)'(\sin \Phi)/\alpha_o - (1 + \varepsilon_1)\delta \Phi,$$

$$\delta \kappa_1 = -(\delta \Phi)'/\alpha_o, \quad \delta \kappa_2 = -\delta \Phi (\cos \Phi)/r_o. \qquad (1.4.3)$$

We substitute (1.4.3) in (1.4.2) and then in (1.4.1). After integration by parts and collecting the coefficients of the independent variations, we obtain

$$-\iint_{O_o} [\{ [r_o(\bar{N}_1^o \sin \Phi + Q^o \cos \Phi)]' + \alpha_o r_o (1 + \varepsilon_1)(1 + \varepsilon_2)p_w \} \delta w$$

$$+ \{ [r_o(\bar{N}_1^o \cos \Phi - Q^o \sin \Phi)]' - \alpha_o \bar{N}_2^o + (1 + \varepsilon_1)(1 + \varepsilon_2)p_u \} \delta u$$

$$- \{[r_o M_1^o]' - \alpha_o M_2^o \cos \Phi + \alpha_o r_o (\gamma_o \bar{N}_1^o - Q^o)\} \delta \Phi] d\xi d\theta$$
$$+ [r_o(\bar{N}_1^o \sin \Phi + Q^o \cos \Phi) \delta w + r_o(\bar{N}_1^o \cos \Phi$$
$$- Q^o \sin \Phi) \delta u - r_o M_1^o \delta \Phi]|_{\xi=\xi_0}^{\xi=\xi_1} = 0. \tag{1.4.4}$$

Due to the independence of variations δw, δu, $\delta \Phi$, from (1.4.4) the equilibrium equations follow

$$(r_o \bar{V}^o)' + \alpha_o r_o p_w^o = 0,$$
$$(r_o \bar{H}^o)' - \alpha_o \bar{N}_2^o + \alpha_o r_o p_u^o = 0,$$
$$(r_o M_1^o)' - \alpha_o M_2^o \cos \Phi + \alpha_o r_o [\gamma_o \bar{N}_1^o - (1 + \varepsilon_1) Q^o] = 0, \tag{1.4.5}$$

and boundary conditions also follow

$$(r_o \bar{V}^o \delta w + r_o \bar{H}^o \delta u - r_o M_1^o \delta \Phi)\big|_{\xi=\xi_b}^{\xi=\xi_e} = 0, \tag{1.4.6}$$

where

$$\bar{V}^o = \bar{N}_1^o \sin \Phi + Q^o \cos \Phi,$$
$$\bar{H}^o = \bar{N}_1^o \cos \Phi - Q^o \sin \Phi,$$
$$p_w^o = (1 + \varepsilon_1)(1 + \varepsilon_2) p_w,$$
$$p_u^o = (1 + \varepsilon_1)(1 + \varepsilon_2) p_u. \tag{1.4.7}$$

Here, $\bar{V}^o$ and $\bar{H}^o$ have the meaning of the internal forces oriented, respectively, vertically (along the symmetry axis) and horizontally (along the radius of the cylindrical coordinate system).

1.5. On the Choice of the Elasticity Ratio

As it is known, equations of equilibrium and conditions of compatibility of deformations alone are not enough to solve the problems of shell theory. When the resolving system of equations is closed, relations that determine the physical properties of the material are required.

In the nonlinear theory of perfectly elastic bodies [33], the task of elasticity relations (equations of state defining equations) is associated with the requirement of the existence of a stored energy function. It is assumed that the work of the loading forces is spent on the message to the body of potential energy, which is returned without loss during unloading (loading and unloading are performed quasi-statically). This assumption is consistent with the physical representations of the deformation of elastic bodies. The elasticity relations in this approach either follow from a reasonably assigned dependence of the potential energy on the invariants of the strain tensors (or) measures, or are given as a direct relationship between the stress tensors and the strain tensors (measures) that are coaxial to them. In the latter case, of course, the conditions for the existence of potential energy should be provided.

In the previous sections, the basic equations of the theory of shells under axisymmetric deformation were obtained from the relations of the three-dimensional theory of elasticity on the basis of geometric hypotheses. The application of the hypotheses of the direct normal and Kirchhoff allows, as it is known, to reduce the three-dimensional problem of elasticity theory to the two-dimensional one, which is the task of constructing the theory of shells.

The elasticity relations of the nonlinear shell theory can also be formulated in terms of the integral characteristics of the stress state and the components of the deformation of the middle surface by the transition from the selected equations of state of the three-dimensional elasticity theory. When solving the problem of setting the elasticity relations, the way of constructing the theory of shells as essentially two-dimensional is also possible. This approach is typical for the works of E. Reissner [37, 38, 43, 44]. We use this approach in the reasoning of this section [40].

We will consider the above relations without taking into account the compression of the material by normal. Specific elementary work of internal forces can be written in two forms

$$\widehat{\delta} A_{(i)} = -(N_1^o \delta\varepsilon_1 + N_2^o \delta\varepsilon_2 + Q^o \delta\gamma_o + M_1^o \delta\kappa_1 + M_2^o \delta\kappa_2) \quad (1.5.1a)$$

or

$$\widehat{\delta} A_{(i)} = -[(1 + \varepsilon_2)N_1\delta\varepsilon_1 + (1 + \varepsilon_1)N_2\delta\varepsilon_2 + (1 + \varepsilon_2)Q\delta\gamma_o$$
$$+ (1 + \varepsilon_2)M_1\delta\kappa_1 + (1 + \varepsilon_1)M_2\delta\kappa_2], \quad (1.5.1b)$$

where

$$N_1 = \int_{-h_o/2}^{h_o/2} t_{(11)}(1 - \zeta \bar{k}_2)d\zeta; \quad N_2 = \int_{-h_o/2}^{h_o/2} t_{(22)}(1 - \zeta \bar{k}_1)d\zeta; \quad (1.5.2)$$

$$Q = \int_{-h_o/2}^{h_o/2} t_{(13)}(1 - \zeta \bar{k}_2)d\zeta; \quad (1.5.3)$$

$$M_1 = \int_{-h_o/2}^{h_o/2} t_{(11)}(1 - \zeta \bar{k}_2)\zeta d\zeta; \quad M_2 = \int_{-h_o/2}^{h_o/2} t_{(22)}(1 - \zeta \bar{k}_1)\zeta d\zeta;$$

$$(1.5.4)$$

$$\bar{k}_1 = \Phi'/\alpha, \quad \bar{k}_2 = (\sin \Phi)/r. \quad (1.5.5)$$

Let us explain the meaning of the introduced integral characteristics with and without zeroes at the top. Values N_1, N_2 are membrane forces, Q is shear force, and M_1, M_2 are bending moments per unit length of the coordinate lines of the surface O. The characteristics with zeros refer to the metric of the initial configuration and are analogs of the stresses introduced in the nonlinear elasticity theory by Treffz, Gamel, Kappus, and other authors. The stress tensor $\hat{T}^o$ introduced by them is related to the Cauchy tensor $\hat{T}$ by the relation

$$\hat{T}^o = \sqrt{\frac{G}{G^o}}\hat{T} = \sqrt{\frac{G}{G^o}}t^{sk}\boldsymbol{R}_s\boldsymbol{R}_k = \overset{o}{t}{}^{sk}\boldsymbol{R}_s\boldsymbol{R}_k. \quad (1.5.6)$$

Then, applying the formula (1.3.5) to an arbitrarily oriented platform $Nd\tilde{O}$ in the volume V into which the platform $Nd\tilde{O}_o$ of volume V_o passes, we obtain

$$\overset{N}{t}\frac{d\tilde{O}}{d\tilde{o}} = \overset{o}{t}{}^{sk}\boldsymbol{R}_s\boldsymbol{N}_k^o = \overset{N}{t}{}^o, \quad (1.5.7)$$

where N_k^o are the guiding cosines of the unit vector are normal $\boldsymbol{N}^o$ in the basis $\boldsymbol{R}_k^o$ and $\overset{N}{t}{}^o$ is the stress vector on the oriented site $Nd\tilde{O}$, but calculated per unit area that this site had in the initial state ($\boldsymbol{N}d\tilde{O}_o$ in volume V_o).

Therefore, keeping in mind the formulas (1.3.4)–(1.3.7) and (1.5.2)–(1.5.4), we get

$$N_1^o = \int_{-h_o/2}^{h_o/2} t_{(11)}^o (1 - \zeta k_2^o) d\zeta, \quad N_2^o = \int_{-h_o/2}^{h_o/2} t_{(22)}^o (1 - \zeta k_1^o) d\zeta,$$

$$M_1^o = \int_{-h_o/2}^{h_o/2} t_{(11)}^o (1 - \zeta k_2^o) \zeta d\zeta, \quad M_2^o = \int_{-h_o/2}^{h_o/2} t_{(22)}^o (1 - \zeta k_1^o) \zeta d\zeta,$$

$$Q^o = \int_{-h_o/2}^{h_o/2} t_{(13)}^o (1 - \zeta k_2^o) d\zeta, \qquad (1.5.8)$$

where the values

$$t_{(11)}^o = (1 + \varepsilon_2^\zeta) t_{(11)}, \quad t_{(22)}^o = (1 + \varepsilon_2^\zeta) t_{(22)}, \quad t_{(13)}^o = (1 + \varepsilon_2^\zeta) t_{(13)} \qquad (1.5.9)$$

are the components of the decomposition of vectors $\overset{1}{t}{}^o$, $\overset{2}{t}{}^o$ in normed basis $\boldsymbol{E}_k$

$$\overset{1}{\boldsymbol{t}}{}^o = t_{(11)}^o \boldsymbol{E}_1 + t_{(13)}^o \boldsymbol{E}_3, \quad \overset{2}{\boldsymbol{t}}{}^o = t_{(22)}^o \boldsymbol{E}_2. \qquad (1.5.10)$$

Then, the equilibrium equations can be represented in two forms

$$(rV)' + \alpha r p_w = 0,$$
$$(rH)' - \alpha N_2 + \alpha r p_u = 0,$$
$$(rM_1)' - \alpha M_2 \cos \Phi + \alpha r (\gamma N_1 - Q) = 0, \qquad (1.5.11)$$

where

$$V = N_1 \sin \Phi + Q \cos \Phi,$$
$$H = N_1 \cos \Phi - Q \sin \Phi; \qquad (1.5.12)$$
$$\alpha = \alpha_o(1 + \varepsilon_1), \quad r = r_o(1 + \varepsilon_2),$$
$$\gamma = \gamma_o(1 + \varepsilon_1), \qquad (1.5.13)$$

and

$$(r_o V^o)' + \alpha_o r_o p_w^o = 0,$$
$$(r_o H^o)' - \alpha_o N_2^o + \alpha_o r_o p_u^o = 0,$$
$$(r_o M_1^o)' - \alpha_o M_2^o \cos \Phi + \alpha_o r_o [\gamma_o N_1^o - (1 + \varepsilon_1) Q^o] = 0, \qquad (1.5.14)$$

where

$$N_1^o = (1+\varepsilon_2)N_1, \quad N_2^o = (1+\varepsilon_1)N_2, \quad Q^o = (1+\varepsilon_2)Q,$$
$$M_1^o = (1+\varepsilon_2)M_1, \quad M_2^o = (1+\varepsilon_2)M_2,$$
$$V^o = (1+\varepsilon_2)V, \quad H^o = (1+\varepsilon_2)H; \tag{1.5.15}$$
$$p_w^o = (1+\varepsilon_1)(1+\varepsilon_2)p_w,$$
$$p_u^o = (1+\varepsilon_1)(1+\varepsilon_2)p_u. \tag{1.5.16}$$

Generalized power characteristics in (1.5.11)–(1.5.12) are related to the metric of the current configuration, and in (1.5.14)–(1.5.16) — to the original.

According to the "two-dimensional" approach, we assume that there is an ideal potential energy of the shell $U_{(i)} = -A_{(i)}$ as a function of the components $\varepsilon_1, \varepsilon_2, \gamma_o, \kappa_1, \kappa_2$. The existence conditions $U_{(i)} = U_{(i)}(\varepsilon_1, \varepsilon_2, \gamma_o, \kappa_1, \kappa_2)$ follow from the necessary and sufficient conditions of integrability (1.5.1). Indeed, if $\delta A_{(i)}$ is total differential, then

$$\delta A_{(i)} = \frac{\partial A_{(i)}}{\partial \varepsilon_1}\delta\varepsilon_1 + \frac{\partial A_{(i)}}{\partial \varepsilon_2}\delta\varepsilon_2 + \frac{\partial A_{(i)}}{\partial \gamma_o}\delta\gamma_o + \frac{\partial A_{(i)}}{\partial \kappa_1}\delta\kappa_1 + \frac{\partial A_{(i)}}{\partial \kappa_2}\delta\kappa_2.$$
$$\tag{1.5.17}$$

Independence of the second derivatives from the order of differentiation

$$\frac{\partial^2 A_{(i)}}{\partial\varepsilon_1\partial\varepsilon_2} = \frac{\partial^2 A_{(i)}}{\partial\varepsilon_2\partial\varepsilon_1}, \dots, \frac{\partial^2 A_{(i)}}{\partial\kappa_1\partial\kappa_2} = \frac{\partial^2 A_{(i)}}{\partial\kappa_2\partial\kappa_1}$$

leads to necessary and sufficient conditions of existence $U_{(i)}$

$$\frac{\partial[(1+\varepsilon_2)N_1]}{\partial\varepsilon_2} = \frac{\partial[(1+\varepsilon_1)N_2]}{\partial\varepsilon_1}, \dots, \frac{\partial[(1+\varepsilon_2)M_1]}{\partial\kappa_2}$$
$$= \frac{\partial[(1+\varepsilon_1)M_2]}{\partial\kappa_1}, \tag{1.5.18}$$

or

$$\frac{\partial N_1^o}{\partial\varepsilon_2} = \frac{\partial N_2^o}{\partial\varepsilon_1}, \dots, \frac{\partial M_1^o}{\partial\kappa_2} = \frac{\partial M_2^o}{\partial\kappa_1}. \tag{1.5.19}$$

From (1.5.1), (1.5.17), it follow the formulas

$$N_1^o = \frac{\partial U_{(i)}}{\partial \varepsilon_1}, \quad N_2^o = \frac{\partial U_{(i)}}{\partial \varepsilon_2}, \quad Q^o = \frac{\partial U_{(i)}}{\partial \gamma_o},$$

$$M_1^o = \frac{\partial U_{(i)}}{\partial \kappa_1}, \quad M_2^o = \frac{\partial U_{(i)}}{\partial \kappa_2}. \tag{1.5.20}$$

The expressions (1.5.20) do not differ in form from similar relations of the theory of shells based on the assumption of smallness of relative elongations. Therefore, a natural simple variant of the elasticity relations will be their choice in the same form in which it was used in the above theory [43], namely, in the form of

$$N_1^o = B(\varepsilon_1 + \nu\varepsilon_2), \quad N_2^o = B(\varepsilon_2 + \nu\varepsilon_1), \quad Q^o = C\gamma_o,$$

$$M_1^o = D(\kappa_1 + \nu\kappa_2), \quad M_2^o = D(\kappa_2 + \nu\kappa_1). \tag{1.5.21}$$

It is easy to verify that the ratios (1.5.21) satisfy the potentiality conditions (1.5.19). In contrast to (1.5.21), constitutive equations

$$N_1 = B(\varepsilon_1 + \nu\varepsilon_2), \quad N_2 = B(\varepsilon_2 + \nu\varepsilon_1), \tag{1.5.22}$$

$$M_1 = D(\kappa_1 + \nu\kappa_2), \quad M_2 = D(\kappa_2 + \nu\kappa_1), \tag{1.5.23}$$

or (1.5.22) and

$$M_1 = D(\tilde{\kappa}_1 + \nu\tilde{\kappa}_2), \quad M_2 = D(\tilde{\kappa}_2 + \nu\tilde{\kappa}_1), \tag{1.5.24}$$

where

$$\tilde{\kappa}_1 = \frac{\Phi_o'}{\alpha_o} - \frac{\Phi'}{\alpha}, \quad \tilde{\kappa}_2 = \frac{\sin\Phi_o}{r_o} - \frac{\sin\Phi}{r}, \tag{1.5.25}$$

do not satisfy the ratio of potentiality (1.5.18), which is easy to see. For example, for (1.5.17)

$$\frac{\partial[(1 + \varepsilon_2)N_1]}{\partial \varepsilon_2} = B(\nu + \varepsilon_1 + 2\nu\varepsilon_2), \tag{1.5.26}$$

$$\frac{\partial[(1 + \varepsilon_1)N_2]}{\partial \varepsilon_1} = B(\nu + \varepsilon_2 + 2\nu\varepsilon_1). \tag{1.5.27}$$

The ratio of (1.5.22)–(1.5.24) has been used in the works [1, 34, 57, 61], with the aim of clarifying governing equations of E. Reissner [41–43]. The main point of refinement is to use the equilibrium equations in the form of (1.5.11) (at $\gamma = 0$), i.e. taking into account changes in the metric of the middle surface. However, these equations can be interpreted as another form of equations (1.5.12), if the forces, moments, and load terms with zeros are entered by formulas (1.5.8), (1.5.15) and (1.5.16). However, the form of the equation (1.5.14) coincides with the equilibrium equations [42, 43]. Thus, if in the equilibrium equations [42, 43], the integral characteristics and the load are calculated in the metric of the deformed shell but attributed to the unit length of the original middle surface, then the equations [42, 43] are also true for large relative elongations. Since the compatibility conditions in [1, 34, 57, 61] are the same as those of E. Reissner, the question of refining the equations of E. Reissner is reduced to the refinement of the elasticity relations. But the choice of elasticity relations in the mentioned works is not justified. As shown above, for large elongations, the ratios (1.5.22)–(1.5.24) do not satisfy the conditions for the existence of the potential energy of the shell. Practically, this means that the work of the external forces expended on the deformation of a sheath, generally speaking, is not returned completely when the load is removed and the shell is restored to its the original configuration, i.e. there is a dissipation of energy. This pattern does not correspond to the behavior of perfectly elastic materials under quasi-static loading and unloading. Therefore, the systems of resolving equations proposed in [1, 25, 27, 28] cannot be considered preferable to the equations of E. Reissner. As a rule, they only complicate these equations (see [1, 27]).

It is appropriate to note that E. Reissner limited himself to the consideration of small elongations because in the case of finite elongations the validity of elasticity relations in the form of (1.5.22), (1.5.23) is doubtful (p. 30 in [42]).

The shell material defined by the law (1.5.21) will be called "semi-linear", following A.I. Lurie [33]. The fact is that in the absence of a shift, the directions $\boldsymbol{E}_1$, $\boldsymbol{E}_2$, $\boldsymbol{n} = \boldsymbol{e}_3$ are the main ones for $\hat{G}^\times$, and in this case (1.5.16) (without $Q^o = C\gamma_o$) can be obtained from the corresponding representation of the specific potential energy for the "semi-linear" material in the nonlinear elasticity theory. On the other hand, the relations (1.5.21) coincide in form with the linear elasticity relations of the theory of shells of small relative elongations of E. Reissner.

In expressions for stiffness

$$B = \frac{Eh}{1 - \nu^2}, \quad D = \frac{Eh^3}{12(1 - \nu^2)}, \quad C = E_{13}h \qquad (1.5.28)$$

by constants E, ν, E_{13} we mean known constants of elasticity theory — Young's modulus, Poisson's ratio, shear modulus.

When the ratio of

$$E_{13} = \frac{E}{2(1 + \nu)} \qquad (1.5.29)$$

the shell is isotropic. If (1.5.29) is not performed, the shell is called transversely isotropic [23].

The expression of the specific energy of deformation of a "semi-linear" material through the deformation components has the form

$$U_{(i)} = \frac{1}{2}[B(\varepsilon_1^2 + 2\nu\varepsilon_1\varepsilon_2 + \varepsilon_2^2) + C\gamma_o$$

$$+ D(\kappa_1^2 + 2\nu\kappa_1\kappa_2 + \kappa_2^2)]. \qquad (1.5.30)$$

The questions related to the choice of adequate defining (physical) relations in the problems of large plastic shell shape changes are discussed in Chapter 2.

Chapter 2

Large Plastic Deformation
of Shells of Rotation

The questions of changing the shape of the shells of rotation are relevant both in connection with the technology of plastic molding and the need to develop methods for modeling such processes. It is difficult to cover the diversity of such tasks. Here, we limit ourselves to the tasks of free drawing uniform (hydraulic) pressure in combination with local loads. Mathematical modeling of such problems is relevant for the technology of plastic forming of domed shells having a given critical load of buckling under loading from the convex side. In systems of protection of tanks and valves, working under pressure, from destruction, they are called flapping safety membranes. The membrane bursts when the pressure rises abnormally before the main structure can collapse. Forming of such shells is carried out by free drawing pressure from flat round plates, rigidly pinched on a circular contour. To obtain membranes of high accuracy of operation, an additional effect of the local load of the backpressure in the apex of the dome is used, which corrects the shape of the shell in the desired direction. This adjustment is called artification. The final form of the artificated shell provides a given external pressure load and the initial axisymmetric form of buckling.

Such membranes are used in fast neutron reactor protection systems. Actually, the technology of manufacturing real membranes in the energy sector gave impetus to the development of a mathematical model and method of calculation of such objects. The simulation task includes the forming stage, stability analysis of the obtained shell, and comparison with experimental data.

The understanding that the resulting dome-shaped shell differs from the spherical one during extraction was formed by the authors in earlier works. The expected geometry and its approximation in the class of ellipsoidal shells were considered in [11]. Attention was also drawn to the possibility of a simple approximation of the thickness of the resulting shell, proposed in [30] based on the analysis of the parameters of the technological process of drawing membranes from corrosion-resistant steel 12X18H10T. The work [30] is connected with the technology of manufacturing safety membranes. It contains tables of the average values of the forming pressure for the given heights, thickness distribution, and curvature of the membranes. Extraction was carried out up to the rupture of the samples. The thinning of the shell in the pole was measured. On the basis of measurements, the change in the thickness of the membrane along the meridian during the formation was well approximated by the quadratic power dependence.

The correctness of the representations about the geometry of the dome-shaped shell was confirmed by a later work [39], in which in the terminology of the concepts of elongation and flatness and the relationship between the geometric parameters and the membrane stresses of a non-momentary thin shell loaded with pressure were studied. Elongation or flatness is considered as a measure of the deviation of the body shape from the sphere. It is noted that the change of this value is complex for a thin shell. It is found that for the metals tested in the work, the surface can be considered spherical only in the vicinity of the pole, and, in general, it is either an elongated or a flattened spheroid. It is also noted that in the processing of metals and polymers supplied in the form of a sheet, the method of drawing pressure is used to determine the index of deformability. This indicator is taken as the maximum height of the dome to be formed, at which no destruction has yet occurred.

These *a priori* representations made it possible to apply a semi-inverse solution method. In [17], the analytical solution of the problem of plastic drawing of a spherical dome from a round plate was constructed. Equations are used that take into account large displacements and rotation angles, changes in the metric, compression of the material normal, diagrams of the hardening material, and deformation physical relations with logarithmic elongations. It is shown that in order to ensure the sphericity of the form, in addition to the main uniform pressure, it is necessary to apply the force concentrated at the top, which appears as a constant when integrating the equations.

In [17], the method [16] is generalized in a mathematical model simulating the technology of plastic forming of flapping articulated safety membranes [14,15]. The manufacture of the shell is performed by the extractor pressure from a flat circular plate, clamped along the contour, in combination with an additional force effects in the top, which provides the artified form like an oblate spheroid. The mathematical model [16] was tested in [19], where a comparison between theory and experiment was given, showing their good agreement. These issues are discussed in Chapter 4.

2.1. Approximation of Material Properties

The basic equations of large deformations of shells of rotation, used to construct the plastic forming problem, are presented in Chapter 1 in terms of kinematic relations and equations of equilibrium. The selection of adequate constitutive relations is discussed in the following.

In the formulation of problems on the behavior of materials and structures, beyond the linear elasticity an important component in the construction of models is the choice of relations that determine the physical properties of the material and link the generalized internal forces and deformations.

Physical dependence $\sigma(\bar{e})$ between stress intensity σ and true (logarithmic) strain intensity $\bar{e}$ is widely used to describe material properties in the deformation theory of plasticity. Its application allows us to consider nonlinear stress–strain states at large deformations. Experiments for metals show the proximity of such dependencies obtained under simple (one-parameter) loading. Usually it is either stretching, compression, or torsion of the samples.

Next, we assume that the conditions of simple loading or close to simple are satisfied. A secant module $E_c(\bar{e}) = \sigma(\bar{e})/\bar{e}$ is introduced, so that $\sigma(\bar{e}) = E_c(\bar{e})\bar{e}$ under active loading, when the stress intensity increases at all points of the material; the deformation plastic model is equivalent to some nonlinear elastic material. During unloading, the Bausinger effect and linearly elastic behavior of the material take place.

In the general case of a three-dimensional state, the stress intensity and strain intensity are related to the stress–strain state components (SSS) by formulas

$$\sigma = \frac{1}{\sqrt{2}}\sqrt{(\sigma_{11} - \sigma_{22})^2 + (\sigma_{11} - \sigma_{33})^2 + (\sigma_{22} - \sigma_{33})^2 + 6(\tau_{12}^2 + \tau_{13}^2 + \tau_{23}^2)},$$

$$\bar{e} = \frac{\sqrt{2}}{3}\sqrt{(\bar{e}_{11} - \bar{e}_{22})^2 + (\bar{e}_{11} - \bar{e}_{33})^2 + (\bar{e}_{22} - \bar{e}_{33})^2 + \frac{3}{2}(\gamma_{12}^2 + \gamma_{13}^2 + \gamma_{23}^2)},$$

$$(2.1.1)$$

where

$$\bar{e}_{kk} = \ln(1 + e_{kk}), \quad k = 1, 2, 3. \qquad (2.1.2)$$

The stress and strain components here correspond to the physical ones in orthogonal coordinate systems ($\sigma_{ii} = t_{(ii)}$). In (2.1.1), the shear deformations are assumed to be small compared to unity. For small elongations $\ln(1 + e_{kk}) \approx e_{kk}$, and dashes in (2.1.1) can be removed.

There are different ways to approximate the dependence $\sigma(\bar{e})$ for material classes. Typically, the determination of the constants of the approximations requires a full chart of tensile, compressive, or torsion data. However, an earlier work [24] considered an approach that is rational and convenient for practical application. It allows you to build highly versatile approximations without the need to have complete material test charts.

When determining the type of approximating function, it is assumed to be linear in the elastic region and different in the plastic region, while maintaining the alignment of the main directions of stress and strain deviators. In practice, bilinear, power, and linear-power functions are the most widely used for analytical recording of experimental diagrams on dimensional planes

$$\sigma = E\bar{e}, \bar{e} \le \bar{e}_{0.2}; \sigma = \sigma_{0.2} + \Pi(\bar{e} - \bar{e}_{0.2}), \quad \bar{e} > \bar{e}_{0.2}; \quad (2.1.3)$$

$$\sigma = C\bar{e}^{\eta}, \quad \bar{e} > 0; \qquad (2.1.4)$$

$$\sigma = E\bar{e}, \quad \bar{e} \le \bar{e}_{0.2}; \quad \sigma = C\bar{e}^{\eta}, \quad \bar{e} > \bar{e}_{0.2}, \qquad (2.1.5)$$

where Π, C, η are constants for the material; $\sigma_{0.2}, \bar{e}_{0.2}$ are the stress and strain of the conditional limit of proportionality (or fluidity); and E is Young's modulus.

It is known that the approximation (2.1.3) gives significant errors, when $\Pi = const$. Closer to the description of real diagrams is the group of formulas (2.1.5), which distinguishes the elastic deformation area. The power function of the parabolic type (2.1.4) accurately reflects the experimental curves of a number of materials at the hardening site but does not work in the elastic deformation region. For real materials, $\eta < 1$, so the secant modulus goes to infinity when $\bar{e} = 0$. For a valid chart, they must be the same Young's modulus. However, with large plastic deformations, elastic deformations are negligible, and then such an approximation is quite acceptable. In the theory of small elastic–plastic deformations A.A. Ilyushin proved that it is the power-law approximation of the hardening curve that corresponds to simple loading. This is called the simple loading theorem [34].

Constants C, η in (2.1.4) are determined from the conditions of passage of the approximating curve through the origin and the point of the conditional yield strength or proportionality $(\sigma_{02}, \bar{e}_{02})$ and tensile strength $(\sigma_l, \bar{e}_l)$, where $\bar{e}_{02} = \sigma_{02}/E$:

$$\sigma_{02} = C(\bar{e}_{02})^\eta, \quad \sigma_l = C(\bar{e}_l)^\eta. \qquad (2.1.6)$$

We shall logarithm (2.1.6) on decimal base

$$\lg(\sigma_{02}) = \lg C + \eta \lg(\bar{e}_{02}), \quad \lg l(\sigma_l) = \lg C + \eta \lg(\bar{e}_l). \qquad (2.1.7)$$

Hence,

$$\eta = (\lg \sigma_l - \lg \sigma_{02})/(\lg \bar{e}_l - \lg \bar{e}_{02}), \quad C = \sigma_l/(\bar{e}_l)^\eta. \qquad (2.1.8)$$

The calculation of constants using formulas (2.1.8) allows us to solve problems without constructing complete tensile diagrams. Data on the conditional yield strength and strength for specific materials can be taken from the reference literature.

Other approximations were proposed and discussed in [24], using, along with the mentioned characteristic points, another additional point of the loading diagram. In this work, it is also proposed to translate the description of the properties of materials in the coordinates of dimensionless stresses $\tilde{\sigma} = \sigma/\sigma_l$ and relative true deformations $\tilde{\bar{e}} = \bar{e}/\bar{e}_l$, where $\sigma_l, \bar{e}_l$ are limit stresses and strains.

Coordinates $\tilde{\sigma}$ and $\tilde{\bar{e}}$ form a dimensionless plane, which describes the process of plastic stretching and reflects the principles of similarity. The translation charts of tests of various materials onto this plane gives the division into groups, determining the properties of closely related materials. The groups differ in the shape of the curves and their position, but within the groups the dispersion of the hardening curves of materials is insignificant. Thus, according to [24], the construction of stretching diagrams on the plane $\tilde{\sigma} - \tilde{\bar{e}}$ reveals the possibility of grouping materials. The materials that make up the group differ in the degree of plasticity. The first group corresponds to very low-plastic spring-type steels. The second group includes low-plastic low-strength titanium alloys. The third group includes materials of medium ductility such as aluminum alloys and heat-resistant alloy steels of medium strength. The fourth group is formed by plastic cast bronze. The fifth group corresponds to austenitic stainless high-plastic steel.

It is obvious that the main coefficient of similarity of materials on the unit plane at the power approximation of the hardening diagram is the ratio of the tensile strength to the yield strength, i.e. σ_l/σ_{02}.

For the power approximation of the hardening curve

$$\sigma(\bar{e}) = C\bar{e}^{\eta}, \quad \sigma(\bar{e}) = E_s(\bar{e})\bar{e}, \quad E_s(\bar{e}) = C\bar{e}^{\eta-1}, \qquad (2.1.9)$$

we obtain the following defining functions in the transition to the unit plane

$$\tilde{\sigma} = \sigma/\sigma_l, \quad \bar{e} = \bar{e}_l\tilde{\bar{e}}, \quad \tilde{\sigma}(\tilde{\bar{e}}) = \tilde{C}\bar{e}_l^{\eta}\tilde{\bar{e}}^{\eta} = \tilde{C}_l\tilde{\bar{e}}^{\eta},$$

$$\tilde{C} = \tilde{C}/\sigma_l, \quad \tilde{C}_l = \tilde{C}\bar{e}_l^{\eta} = 1,$$

$$\tilde{E}_s(\tilde{\bar{e}}) = \tilde{C}\bar{e}_l^{(\eta-1)}\tilde{\bar{e}}^{(\eta-1)} = \tilde{C}_E\tilde{\bar{e}}^{(\eta-1)}, \quad \tilde{C}_E = \tilde{C}\bar{e}_l^{(\eta-1)} = 1/\bar{e}_l. \,(2.1.10)$$

In the calculations, we will consider materials such as stainless steels 09X18H9, 12H18N9, and 12H18N10T, which have similar characteristics. So, according to [32], for the material 12X18H10T: $E = 0.21 \cdot 10^6\,MPa$, $\sigma_{0.2} = 360\,MPa$, $\bar{e}_{0.2} = \sigma_{0.2}/E = 0.001714$, $\sigma_l = 720\,MPa$, $\bar{e}_l = 0.615$.

Power approximation constants have the following values: $\eta = 0.1178256$, $C = 762.445\,MPa$.

2.2. Nonlinear Physical Relations

In addition to information about the properties of the material for solving physically nonlinear problems, it is necessary to have defining relations connecting the components of stresses and strains. Relatively simple resolving equations with logarithmic deformations are formed on the basis of physical ratios for incompressible materials proposed by Davis and Nadai [31, 35] (DN ratios). We consider the normal stresses to coincide with the principal and oriented along the coordinate directions. In the three-dimensional version of the ratio of the DN for $\bar{e} \geq \bar{e}_{0.2}$, we write in the form

$$E_c \bar{e}_{11} = \sigma_{11} - 0.5(\sigma_{22} + \sigma_{33}),$$

$$E_c \bar{e}_{22} = \sigma_{22} - 0.5(\sigma_{11} + \sigma_{33}),$$

$$E_s \bar{e}_{33} = \sigma_{33} - 0.5(\sigma_{11} + \sigma_{22}). \tag{2.2.1}$$

Formally, the entry (2.2.1) is similar to Hooke's law. Only here, the deformations are understood as logarithmic deformations (true, natural), and Young's modulus is replaced by a secant modulus depending on the intensity of the deformations and, accordingly, on the coordinates of the current point.

In the theory of shells, stresses σ_{33} on equidistant surfaces in the thickness of the shell are neglected in comparison with the rest ($\sigma_{33} \approx 0$). Next, we consider the axisymmetric stress–strain state, determined by the components

$$\sigma_{11} = \sigma_1, \quad \sigma_{22} = \sigma_2, \quad \sigma_{13} = \tau_{13}, \quad \bar{e}_{11} = \bar{e}_1,$$

$$\bar{e}_{22} = \bar{e}_2, \quad e_{13} = 0.5\gamma_{13} = 0.5\gamma.$$

Since $\gamma_{13} = \gamma$ is assumed to be small, the components of normal stresses and strains are close to the main directions coinciding with the coordinate ones. For the ratio of the bottom of the shear angle, γ is generally assumed to be zero.

When $\sigma_{33} = 0$ from (2.2.1), it follows that

$$\sigma_1 = (3E_s/4)(\bar{e}_1 + 0.5\bar{e}_2), \quad \sigma_2 = (3E_s/4)(\bar{e}_2 + 0.5\bar{e}_1); \tag{2.2.2}$$

$$\bar{e}_3 = -(\bar{e}_1 + \bar{e}_2). \tag{2.2.3}$$

On the middle surface ($\zeta = 0$)

$$\bar{\varepsilon}_3 = -(\bar{\varepsilon}_1 + \bar{\varepsilon}_2). \tag{2.2.4}$$

Ratios (2.2.3) and (2.2.4) are the conditions of incompressibility of the material. For an incompressible material and $\gamma = 0$, recording the intensity of deformation (2.1.1) is simplified

$$\bar{e} = (2/\sqrt{3})\sqrt{\bar{e}_1^2 + \bar{e}_1\bar{e}_2 + \bar{e}_2^2},$$

$$\bar{\varepsilon} = (2/\sqrt{3})\sqrt{\bar{\varepsilon}_1^2 + \bar{\varepsilon}_1\bar{\varepsilon}_2 + \bar{\varepsilon}_2^2}. \tag{2.2.5}$$

Let us consider the defining relations of the theory of small elasto-plastic deformations. This theory, proposed by Genki and significantly developed in the works of A.A. Ilyushin, is based on the following hypotheses [29, 34]:

(1) The volume deformation is proportional to the average normal stress with the same coefficient of proportionality as within the elasticity, i.e. due to the plastic deformation, there is no change in the volume.
(2) The components of the deviator of the deformation are proportional to the components of the stress deviator.
(3) Stress intensity is a function of strain intensity independent of the type of stress state.

The corresponding equations can be written as

$$e_{11} - e_0 = (3/2\Lambda)(\sigma_{11} - \sigma_0),$$

$$e_{22} - e_0 = (3/2\Lambda)(\sigma_{22} - \sigma_0),$$

$$e_{33} - e_0 = (3/2\Lambda)(\sigma_{33} - \sigma_0); \tag{2.2.6}$$

$$\gamma_{12} = (3/\Lambda)\tau_{12}, \quad \gamma_{13} = (3/\Lambda)\tau_{13},$$

$$\gamma_{23} = (3/\Lambda)\tau_{23}, \tag{2.2.7}$$

where $e_0 = \Theta/3$, $\Theta = e_{11} + e_{22} + e_{33}$ is the volumetric deformation, $\sigma_0 = (\sigma_{11} + \sigma_{22} + \sigma_{33})/3$ is the octahedral ("average" normal) stress; $\Lambda = E$ by $e < e_{02}$ (in areas of elasticity), $\Lambda = E_s(e)$ by $e > e_{02}$ (in the plastic zones).

Values Θ and σ_0 are related through the bulk modulus of elasticity

$$\Theta = \sigma_0/K, \quad K = E/[3(1 - 2\nu)]. \tag{2.2.8}$$

Relations (2.2.6) can also be given the form of a generalized Hooke's law by substituting expressions e_0 and σ_0. After the transformations, we obtain

$$e_{11} = (E_p)^{-1}[\sigma_{11} - \nu_p(\sigma_{22} + \sigma_{33})],$$
$$e_{22} = (E_p)^{-1}[\sigma_{22} - \nu_p(\sigma_{11} + \sigma_{33})],$$
$$e_{33} = (E_p)^{-1}[\sigma_{33} - \nu_p(\sigma_{11} + \sigma_{22})], \qquad (2.2.9)$$

where

$$E_p = \frac{\Lambda}{1 + \Lambda/(9K)}, \quad \nu_p = \frac{0.5 - \Lambda/(9K)}{1 + \Lambda/(9K)}, \quad G_p = \Lambda/3. \qquad (2.2.10)$$

If the physical equations of the theory of small elastic–plastic deformations are taken in the form (2.2.9), then the solution of plasticity problems is reduced to the solution of problems of the theory of elasticity with variable parameters of elasticity. In this case, the relationship between K_p (volume modulus) and G_p (shear modulus) with E_p and ν_p has the same form as for elastic constants

$$K_p = \frac{E_p}{3(1 - 2\nu_p)}, \quad G_p = \frac{E_p}{2(1 + \nu_p)}. \qquad (2.2.11)$$

Such interpretation is offered by A.A. Ilyushin. The iterative method of variable parameters of elasticity, based on this, was developed by I.A. Birger [36]. In the case of an incompressible material, $\nu \to 0.5$, $K \to \infty$, $E_p = 3G_p \to \Lambda$, $\nu_p \to 0.5$.

With regard to the considered stress–strain state for shells from (2.2.9), it follows that

$$\sigma_1 = \frac{E_p}{1 - \nu_p^2}(e_1 + \nu_p e_2), \quad \sigma_2 = \frac{E_p}{1 - \nu_p^2}(e_1 + \nu_p e_2); \qquad (2.2.12)$$

$$e_3 = \frac{-\nu_p}{1 - \nu_p}(e_1 + e_2). \qquad (2.2.13)$$

On the middle surface

$$\varepsilon_3 = \frac{-\nu_p}{1 - \nu_p}(\varepsilon_1 + \varepsilon_2). \qquad (2.2.14)$$

Analysis of the behavior of the coefficient of transverse deformation with increasing strain intensity in the plastic region shows that it very quickly tends to 0.5. This is confirmed experimentally [34]. Already at one percent of the strain intensity, the Poisson's ratio reaches the value $\nu_p = 0.48$ at the initial elastic $\nu = 0.3$. Therefore, the assumption of incompressibility of the material is quite acceptable for applied problems.

In the technical literature related to the development of engineering models for the calculation of stamped products from thin-walled billets, there are statements about the efficiency of the theory of small elastoplastic deformations and in the field of sufficiently developed (large) deformations. So, numerical experiments based on the relations (2.2.12)–(2.2.14) taking into account changes in the metric during deformation make sense.

On the basis of (2.2.14), the variation of transverse deformation is

$$\delta\varepsilon_3 = \frac{-\nu_p}{1-\nu_p}(\delta\varepsilon_1 + \delta\varepsilon_2). \tag{2.2.15}$$

In this embodiment, in the expression of virtual work (1.3.18) under the generalized effort $\bar{N}_1^o$ and $\bar{N}_2^o$, we should also understand that

$$\bar{N}_1^o = N_1^o + \frac{\nu_p}{(1-\nu_p)(1+\varepsilon_3)}(K_1 M_1^o + K_2 M_2^o), \tag{2.2.16}$$

$$\bar{N}_2^o = N_2^o + \frac{\nu_p}{(1-\nu_p)(1+\varepsilon_3)}(K_1 M_1^o + K_2 M_2^o). \tag{2.2.17}$$

In principle, the value of ν_p depends on the deformation, but it makes no sense to change it due to the rapid output of almost a constant in the development of plastic deformations, for which it makes sense to correct N_1^o and N_2^o in (2.2.12) and (2.2.14).

In the case of the transition to the logarithmic strains in (2.2.12)–(2.2.14),

$$\sigma_1 = \frac{E_p}{1-\nu_p^2}(\bar{e}_1 + \nu\bar{e}_2), \quad \sigma_2 = \frac{E_p}{1-\nu_p^2}(\bar{e}_2 + \nu\bar{e}_1); \tag{2.2.18}$$

$$\bar{e}_3 = \frac{-\nu_p}{1-\nu_p}(\bar{e}_1 + \bar{e}_2); \tag{2.2.19}$$

$$\frac{-\delta\varepsilon_3}{1+\varepsilon_3} = \frac{\nu_p}{1-\nu_p}\left(\frac{\delta\varepsilon_1}{1+\varepsilon_1} + \frac{\delta\varepsilon_2}{1+\varepsilon_2}\right), \tag{2.2.20}$$

$$\bar{N}_1^o = N_1^o + \frac{\nu_p}{(1 - \nu_p)(1 + \varepsilon_1)}(K_1 M_1^o + K_2 M_2^o), \quad (2.2.21)$$

$$\bar{N}_2^o = N_2^o + \frac{\nu_p}{(1 - \nu_p)(1 + \varepsilon_2)}(K_1 M_1^o + K_2 M_2^o). \quad (2.2.22)$$

The relatively simple physical relations used in the problems of the nonlinear elasticity theory include the defining equations of the "semi-linear" (harmonic) material, connecting the main stresses and the main relative elongations [33]. With respect to the radial axisymmetric deformation of the shells of rotation, taking into account $e_3 = \varepsilon_3$, these relations are written in the form

$$\sigma_1 = \frac{1 + e_1}{\sqrt{I_3}} \cdot \frac{E}{1 - \nu^2}(e_1 + \nu e_2),$$

$$\sigma_2 = \frac{1 + e_2}{\sqrt{I_3}} \cdot \frac{E}{1 - \nu^2}(e_2 + \nu e_1), \quad (2.2.23)$$

where I_3 is the third invariant of the first measure of deformation,

$$\sqrt{I_3} = (1 + e_1)(1 + e_2)(1 + \varepsilon_3), \quad \varepsilon_3 = -\nu(1 - \nu)^{-1}(\varepsilon_1 + \varepsilon_2). \quad (2.2.24)$$

Taking into account (2.2.23),

$$\sigma_1 = \frac{1}{(1 + e_2)(1 + \varepsilon_3)} \cdot \frac{E}{1 - \nu^2}(e_1 + \nu e_2),$$

$$\sigma_2 = \frac{1}{(1 + e_1)(1 + \varepsilon_3)} \cdot \frac{E}{1 - \nu^2}(e_2 + \nu e_1). \quad (2.2.25)$$

Following the logic of constructing the above relations for modeling plastic deformation, the module E in (2.2.25) can be replaced by Λ or E_p, ν by 0.5 or ν_p. Further generalization, by analogy with the relations of DN, is associated with the replacement of relative elongations by logarithmic deformations.

As further constructions show, the use of relations of the type (2.2.25) leads to the simplest connections of the integral characteristics of the stress state, introduced by formulas (1.3.12), with generalized deformations.

Indeed, for the efforts and moments we have

$$N_1^o = (1 + \varepsilon_3) \int_{-h_o/2}^{h_o/2} \sigma_1(1 + e_2)d\zeta,$$

$$N_2^o = (1 + \varepsilon_3) \int_{-h_o/2}^{h_o/2} \sigma_2(1 + e_1)d\zeta; \qquad (2.2.26)$$

$$M_1^o = (1 + \varepsilon_3)^2 \int_{-h_o/2}^{h_o/2} \sigma_1(1 + e_2)\zeta d\zeta,$$

$$M_2^o = (1 + \varepsilon_3)^2 \int_{-h_o/2}^{h_o/2} \sigma_2(1+e_1)\zeta d\zeta, \qquad (2.2.27)$$

where $t_{(11)} = \sigma_1$, $t_{(22)} = \sigma_2$, $\varepsilon_1^\zeta = e_1$, $\varepsilon_2^\zeta = e_2$.

Substituting (2.2.25) with modules of plasticity in (2.2.26), (2.2.27) and the transition to an incompressible material gives

$$N_1^o = \frac{4}{3} \int_{-h_o/2}^{h_o/2} \Lambda(e_1 + 0.5e_2)_1 d\zeta,$$

$$N_2^o = \frac{4}{3} \int_{-h_o/2}^{h_o/2} \Lambda(e_2 + 0.5e_1)_2 d\zeta; \qquad (2.2.28)$$

$$M_1^o = \frac{4}{3}(1 + \varepsilon_3) \int_{-h_o/2}^{h_o/2} \Lambda(e_1 + 0.5e_2)_1 \zeta d\zeta,$$

$$M_2^o = \frac{4}{3}(1 + \varepsilon_3) \int_{-h_o/2}^{h_o/2} \Lambda(e_2 + 0.5e_1)_2 \zeta d\zeta. \qquad (2.2.29)$$

Here, you can replace $(1+\varepsilon_3)$ with the product $(1+\varepsilon_1)^{-1}(1+\varepsilon_2)^{-1}$, when it is convenient, and conversely, replace $(1 + \varepsilon_1)(1 + \varepsilon_2)$ with $(1 + \varepsilon_3)^{-1}$.

For material type DN,

$$N_1^o = \frac{4}{3}\delta_3 \int_{-h_o/2}^{h_o/2} \Lambda(\bar{e}_1 + 0.5\bar{e}_2)_1 \Delta_2 d\zeta,$$

$$N_2^o = \frac{4}{3}\delta_3 \int_{-h_o/2}^{h_o/2} \Lambda(\bar{e}_2 + 0.5\bar{e}_1)_2 \Delta_1 d\zeta; \qquad (2.2.30)$$

$$M_1^o = \frac{4}{3}\delta_3^2 \int_{-h_o/2}^{h_o/2} \Lambda(\bar{e}_1 + 0.5\bar{e}_2)_1 \Delta_2 \zeta d\zeta,$$

$$M_2^o = \frac{4}{3}\delta_3^2 \int_{-h_o/2}^{h_o/2} \Lambda(\bar{e}_2 + 0.5\bar{e}_1)_2 \Delta_1 \zeta d\zeta, \qquad (2.2.31)$$

where $\delta_k = 1 + \varepsilon_k$, $k = 1, 2, 3$.

We will present the logarithmic strain and multipliers Δ_1, and Δ_2 as follows:

$$\bar{e}_1 = \ln(1 + \varepsilon_1 + \zeta\kappa_1) = \ln(1 + \varepsilon_1) + \ln(1 + \zeta\bar{\kappa}_1),$$

$$\bar{e}_2 = \ln(1 + \varepsilon_2 + \zeta\kappa_2) = \ln(1 + \varepsilon_2) + \ln(1 + \zeta\bar{\kappa}_2),$$

$$\Delta_1 = 1 + \varepsilon_1 + \zeta\kappa_1 = \delta_1(1 + \zeta\bar{\kappa}_1),$$

$$\Delta_2 = 1 + \varepsilon_2 + \zeta\kappa_2 = \delta_2(1 + \zeta\bar{\kappa}_2), \qquad (2.2.32)$$

where

$$\bar{\kappa}_1 = \kappa_1/\delta_1, \quad \bar{\kappa}_2 = \kappa_2/\delta_2. \qquad (2.2.33)$$

The values $\bar{\kappa}_1$ and $\bar{\kappa}_2$ have the meaning of changes in the main curvatures of the middle surface of the deformed shell, reduced by $\delta_1 = 1 + \varepsilon_1$ and $\delta_2 = 1 + \varepsilon_2$, respectively, where $\varepsilon_1 > 0$, $\varepsilon_2 > 0$. We believe that the original shell is thin and remains so in the process of deformation. Then, the values $\zeta\bar{\kappa}_1$ and $\zeta\bar{\kappa}_2$ are small in comparison with unity, so that it can be put as

$$\ln(1 + \zeta\bar{\kappa}_1) \approx \zeta\bar{\kappa}_1, \quad \ln(1 + \zeta\bar{\kappa}_2) \approx \zeta\bar{\kappa}_2$$

with an accuracy of, respectively, $0.5\zeta\bar{\kappa}_1$ and $0.5\zeta\bar{\kappa}_2$ compared with unit.

Therefore, there are ratios

$$\bar{e}_1 = \bar{\varepsilon}_1 + \zeta\bar{\kappa}_1, \quad \bar{e}_2 = \bar{\varepsilon}_2 + \zeta\bar{\kappa}_2. \qquad (2.2.34)$$

With the error $\max\{\zeta\bar{\kappa}_1, \zeta\bar{\kappa}_2\}$ compared to the unit, it can also be put as

$$\Delta_1 = 1 + \varepsilon_1, \quad \Delta_2 = 1 + \varepsilon_2. \qquad (2.2.35)$$

Thus, the brackets $(\cdots)_1$ and $(\cdots)_2$ under the integral are linearly dependent on ζ

$$(\cdots)_1 = a_1 + \zeta b_1, \quad (\cdots)_2 = a_2 + \zeta b_2, \qquad (2.2.36)$$

where

$$a_1 = \varepsilon_1 + 0.5\varepsilon_2, \quad a_2 = \varepsilon_2 + 0.5\varepsilon_1,$$

$$b_1 = \kappa_1 + 0.5\kappa_2, \quad b_2 = \kappa_2 + 0.5\kappa_1. \tag{2.2.37}$$

For logarithmic deformations a_j are replaced by $\bar{a}_j$, b_j – by $\bar{b}_j$

$$\bar{a}_1 = \bar{\varepsilon}_1 + 0.5\bar{\varepsilon}_2, \quad \bar{a}_2 = \bar{\varepsilon}_2 + 0.5\bar{\varepsilon}_1,$$

$$\bar{b}_1 = \bar{\kappa}_1 + 0.5\bar{\kappa}_2, \quad \bar{b}_2 = \bar{\kappa}_2 + 0.5\bar{\kappa}_1. \tag{2.2.38}$$

Since Λ depends on the components of the deformation in a complex way, we cannot precisely take the integrals of (2.2.30), (2.2.31) in explicit form. This can be done approximately by replacing the Λ three-term sum of the Taylor (Laurent) power series by ζ in the neighborhood $\zeta = 0$,

$$\Lambda = A_0 + A_1\zeta + A_2\zeta^2. \tag{2.2.39}$$

In the case of moderate elongation,

$$A_0 = \Lambda(\varepsilon) = C\varepsilon^{\eta-1}, \quad A_1 = [\Lambda(e)_{,e}\, e_{,\zeta}]|_{\zeta=0},$$

$$A_2 = 0.5[\Lambda(e)_{,ee}\, e_{,\zeta} + \Lambda(e)_{,e}\, e_{,\zeta\zeta}]|_{\zeta=0},$$

$$\Lambda(e)_{,e}\,|_{\zeta=0} = C(\eta-1)\varepsilon^{\eta-2},$$

$$\Lambda(e)_{,ee}\,|_{\zeta=0} = C(\eta-1)(\eta-2)\varepsilon^{\eta-3},$$

$$e(\zeta)_{,\zeta}\,|_{\zeta=0} = 2\kappa_\varepsilon/(3\varepsilon),$$

$$e(\zeta)_{,\zeta\zeta}\,|_{\zeta=0} = 2\kappa/(3\varepsilon) - 4\kappa_\varepsilon^2/(9\varepsilon^3);$$

$$\varepsilon = (2/\sqrt{3})\sqrt{\varepsilon_1^2 + \varepsilon_1\varepsilon_2 + \varepsilon_2^2};$$

$$\kappa_\varepsilon = 2(a_1\kappa_1 + a_2\kappa_2)$$

$$= 2[\varepsilon_1\kappa_1 + 0.5(\varepsilon_1\kappa_2 + \varepsilon_2\kappa_1) + \varepsilon_2\kappa_2],$$

$$a_1 = \varepsilon_1 + 0.5\varepsilon_2, \quad a_2 = \varepsilon_2 + 0.5\varepsilon_1,$$

$$\kappa = 2(\kappa_1^2 + \kappa_1\kappa_2 + \kappa_2^2) - 2(a_1\kappa_1^2 + a_2\kappa_2^2). \tag{2.2.40}$$

For logarithmic deformations,

$$\bar{A}_0 = \Lambda(\bar{\varepsilon}), \quad \bar{A}_1 = [\Lambda(\bar{e})_{,\bar{e}}\, \bar{e}_{,\zeta}]|_{\zeta=0},$$

$$\bar{A}_2 = 0.5[\Lambda(\bar{e})_{,\bar{e}\bar{e}}\, \bar{e}_{,\zeta} + \Lambda(\bar{e})_{,\bar{e}}\, \bar{e}_{,\zeta\zeta}]|_{\zeta=0},$$

$$\Lambda(\bar{e}),_{\bar{e}}\,|_{\zeta=0} = C(\eta-1)\bar{e}^{\eta-2},$$

$$\Lambda(\bar{e}),_{\bar{e}\bar{e}}\,|_{\zeta=0} = C(\eta-1)(\eta-2)\bar{e}^{\eta-3},$$

$$\bar{e}(\zeta),_{\zeta}\,|_{\zeta=0} = 2\bar{\kappa}_{\varepsilon}/(3\bar{\varepsilon}),$$

$$\bar{e}(\zeta),_{\zeta\zeta}\,|_{\zeta=0} = 2\bar{\kappa}/(3\bar{\varepsilon}) - 4\bar{\kappa}_{\varepsilon}^2/(9\bar{\varepsilon}^3);$$

$$\bar{\varepsilon} = (2/\sqrt{3})\sqrt{\bar{\varepsilon}_1^2 + \bar{\varepsilon}_1\bar{\varepsilon}_2 + \bar{\varepsilon}_2^2};$$

$$\bar{\kappa}_{\varepsilon} = 2(\bar{a}_1\bar{\kappa}_1 + \bar{a}_2\bar{\kappa}_2)$$

$$= 2[\bar{\varepsilon}_1\bar{\kappa}_1 + 0.5(\bar{\varepsilon}_1\bar{\kappa}_2 + \bar{\varepsilon}_2\bar{\kappa}_1) + \bar{\varepsilon}_2\bar{\kappa}_2],$$

$$\bar{\kappa} = 2(\bar{\kappa}_1^2 + \bar{\kappa}_1\bar{\kappa}_2 + \bar{\kappa}_2^2) - 2(\bar{a}_1\bar{\kappa}_1^2 + \bar{a}_2\bar{\kappa}_2^2). \qquad (2.2.41)$$

Thus, for the material of the "semi-linear" type (SL),

$$N_1^o = \frac{4}{3}\int_{-h_o/2}^{h_o/2}(A_0 + A_1\zeta + A_2\zeta^2)(a_1 + \zeta b_1)d\zeta,$$

$$N_2^o = \frac{4}{3}\int_{-h_o/2}^{h_o/2}(A_0 + A_1\zeta + A_2\zeta^2)(a_2 + \zeta b_2)d\zeta,$$

$$M_1^o = \frac{4}{3}\delta_3\int_{-h_o/2}^{h_o/2}(A_0 + A_1\zeta + A_2\zeta^2)(a_1 + \zeta b_1)\zeta d\zeta,$$

$$M_2^o = \frac{4}{3}\delta_3\int_{-h_o/2}^{h_o/2}(A_0 + A_1\zeta + A_2\zeta^2)(a_2 + \zeta b_2)\zeta d\zeta. \qquad (2.2.42)$$

For ratios of type DN,

$$N_1^o = \frac{4}{3}\delta_3\delta_2\int_{-h_o/2}^{h_o/2}(\bar{A}_0 + \bar{A}_1\zeta + \bar{A}_2\zeta^2)(\bar{a}_1 + \zeta\bar{b}_1)(1 + \zeta\bar{\kappa}_2)d\zeta,$$

$$N_2^o = \frac{4}{3}\delta_3\delta_1\int_{-h_o/2}^{h_o/2}(\bar{A}_0 + \bar{A}_1\zeta + \bar{A}_2\zeta^2)(\bar{a}_2 + \zeta\bar{b}_2)(1 + \zeta\bar{\kappa}_1)d\zeta,$$

$$M_1^o = \frac{4}{3}\delta_3^2\delta_2\int_{-h_o/2}^{h_o/2}(\bar{A}_0 + \bar{A}_1\zeta + \bar{A}_2\zeta^2)(\bar{a}_1 + \zeta\bar{b}_1)(1 + \zeta\bar{\kappa}_2)\zeta d\zeta,$$

$$M_2^o = \frac{4}{3}\delta_3^2\delta_1\int_{-h_o/2}^{h_o/2}(\bar{A}_0 + \bar{A}_1\zeta + \bar{A}_2\zeta^2)(\bar{a}_2 + \zeta\bar{b}_2)(1 + \zeta\bar{\kappa}_1)\zeta d\zeta.$$

$$(2.2.43)$$

After integration for SL material, we have

$$N_1^o = \{Ba_1 + D[\Theta_1 b_1 + \Theta_2 a_1]\},$$
$$N_2^o = \{Ba_2 + D[\Theta_1 b_2 + \Theta_2 a_2]\},$$
$$M_1^o = \delta_3 D(b_1 + \Theta_1 a_1),$$
$$M_2^o = \delta_3 D(b_2 + \Theta_1 a_2), \qquad (2.2.44)$$

where

$$\Theta_1 = A_1/A_0, \quad \Theta_2 = A_2/A_0,$$
$$B = (4/3)\Lambda(\varepsilon)h_o, \quad D = (1/9)\Lambda(\varepsilon)h_o^3. \qquad (2.2.45)$$

The determining relations of the type of DN have the form

$$N_1^o = \delta_3\delta_2\{\bar{B}\bar{a}_1 + \bar{D}[\bar{\Theta}_1\bar{b}_1 + \bar{\Theta}_2\bar{a}_1]\},$$
$$N_2^o = \delta_3\delta_2\{\bar{B}\bar{a}_2 + \bar{D}[\bar{\Theta}_1\bar{b}_2 + \bar{\Theta}_2\bar{a}_2]\},$$
$$M_1^o = \delta_3^2\delta_2\bar{D}(\bar{b}_1 + \bar{\Theta}_1\bar{a}_1),$$
$$M_2^o = \delta_3^2\delta_1\bar{D}(\bar{b}_2 + \bar{\Theta}_1\bar{a}_2), \qquad (2.2.46)$$

where

$$\bar{\Theta}_1 = \bar{A}_1/\bar{A}_0, \quad \bar{\Theta}_2 = \bar{A}_2/\bar{A}_0,$$
$$\bar{B} = (4/3)\Lambda(\bar{\varepsilon})h_o, \quad \bar{D} = (1/9)\Lambda(\bar{\varepsilon})h_o^3. \qquad (2.2.47)$$

In the simplest version of the defining ratios can be put $\Lambda(e) \approx A_0 = \Lambda(\varepsilon), \Lambda(\bar{e}) \approx \bar{A}_0 = \Lambda(\bar{\varepsilon})$. This is justified for deep drawing problems and is equivalent to neglecting the variability of material properties in thickness, i.e. replacing them with appropriate properties on the middle surface. Additionally, we assume that $\max\{\zeta\bar{\kappa}_1, \zeta\bar{\kappa}_2\} \ll 1$. As a result, we have the following nonlinear physical relations.

For material type SL

$$N_1^o = B(\varepsilon_1 + 0.5\varepsilon_2),$$
$$N_2^o = B(\varepsilon_2 + 0.5\varepsilon_1); \qquad (2.2.48)$$
$$M_1^o = D_1(\kappa_1 + 0.5\kappa_2),$$
$$M_2^o = D_1(\kappa_2 + 0.5\kappa_1), \qquad (2.2.49)$$

where

$$B = (4/3)\Lambda(\varepsilon)h_o, \quad D_1 = \delta_3 D = (1/9)\Lambda(\varepsilon)h_o^3.$$

For material type DN

$$N_1^o = \bar{B}_1(\bar{\varepsilon}_1 + 0.5\bar{\varepsilon}_2),$$
$$N_2^o = \bar{B}_2(\bar{\varepsilon}_2 + 0.5\bar{\varepsilon}_1);$$
$$M_1^o = \bar{D}_1(\bar{\kappa}_1 + 0.5\bar{\kappa}_2),$$
$$M_2^o = \bar{D}_2(\bar{\kappa}_2 + 0.5\bar{\kappa}_1), \tag{2.2.50}$$

where

$$\bar{\kappa}_1 = \kappa_1/\delta_1, \quad \bar{\kappa}_2 = \kappa_2/\delta_2, \tag{2.2.51}$$
$$\bar{B}_1 = \bar{B}/\delta_1, \quad \bar{B}_2 = \bar{B}/\delta_2,$$
$$\bar{D}_1 = \delta_3^2\delta_2\bar{D}, \quad \bar{D}_2 = \delta_3^2\delta_1\bar{D}. \tag{2.2.52}$$

With the help of two switches m and n, which take values 0 or 1, it is possible to combine the records of the defining relations of the type DN and SL in the variants of logarithmic and conventional (non-logarithmic) relative elongations. It is convenient for comparison of defining relations in numerical experiments

$$N_1^o = (\delta_3\delta_2)^m[n\bar{B}(\bar{\varepsilon}_1 + 0.5\bar{\varepsilon}_2) + (1-n)B(\varepsilon_1 + 0.5\varepsilon_2)],$$
$$N_2^o = (\delta_3\delta_1)^m[n\bar{B}(\bar{\varepsilon}_2 + 0.5\bar{\varepsilon}_1) + (1-n)B(\varepsilon_2 + 0.5\varepsilon_1)],$$
$$M_1^o = \delta_3^{m+1}\delta_2^m[n\bar{D}(\bar{\kappa}_1 + 0.5\bar{\kappa}_2) + (1-n)D(\kappa_1 + 0.5\kappa_2)],$$
$$M_2^o = \delta_3^{m+1}\delta_1^m[n\bar{D}(\bar{\kappa}_2 + 0.5\bar{\kappa}_1) + (1-n)D(\kappa_2 + 0.5\kappa_1)]. \tag{2.2.53}$$

The following more complex version of the defining relations corresponds to the account of the terms with $\Theta_1 = A_1/A_0$ and $\bar{\Theta}_1 = \bar{A}_1/\bar{A}_0$ by $\Theta_2 = 0$ and $\bar{\Theta}_2 = 0$.

Then for material type SL,

$$N_1^o = B(\varepsilon_1 + 0.5\varepsilon_2) + D\Theta_1(\kappa_1 + 0.5\kappa_2),$$
$$N_2^o = B(\varepsilon_2 + 0.5\varepsilon_1) + D\Theta_1(\kappa_2 + 0.5\kappa_1),$$
$$M_1^o = \delta_3 D[\Theta_1(\varepsilon_1 + 0.5\varepsilon_2) + \kappa_1 + 0.5\kappa_2],$$
$$M_2^o = \delta_3 D[\Theta_1(\varepsilon_2 + 0.5\varepsilon_1) + \kappa_2 + 0.5\kappa_1]. \tag{2.2.54}$$

Material type DN

$$N_1^o = \delta_3\delta_2[(\bar{B} + \bar{D}\bar{\Theta}_1\bar{\kappa}_2)(\bar{\varepsilon}_1 + 0.5\bar{\varepsilon}_2) + \bar{D}(\bar{\Theta}_1 + \bar{\kappa}_2)(\bar{\kappa}_1 + 0.5\bar{\kappa}_2)],$$
$$N_2^o = \delta_3\delta_1[(\bar{B} + \bar{D}\bar{\Theta}_1\bar{\kappa}_1)(\bar{\varepsilon}_2 + 0.5\bar{\varepsilon}_1) + \bar{D}(\bar{\Theta}_1 + \bar{\kappa}_1)(\bar{\kappa}_2 + 0.5\bar{\kappa}_1)],$$

$$M_1^o = \delta_3^2 \delta_2 \bar{D}[(\bar{\Theta}_1 + \bar{\kappa}_2)(\bar{\varepsilon}_1 + 0.5\bar{\varepsilon}_2) + \bar{\kappa}_1 + 0.5\bar{\kappa}_2],$$

$$M_2^o = \delta_3^2 \delta_1 \bar{D}[(\bar{\Theta}_1 + \bar{\kappa}_1)(\bar{\varepsilon}_2 + 0.5\bar{\varepsilon}_1) + \bar{\kappa}_2 + 0.5\bar{\kappa}_1]. \tag{2.2.55}$$

In integrands (2.2.43), the values can be neglected ($\zeta\bar{\kappa}_1$, $\zeta\bar{\kappa}_2$) in comparison with the unit, which somewhat simplifies the option of ratios such as DN. In this case, the relations (2.2.55) convert to the following:

$$N_1^o = \delta_3 \delta_2 [\bar{B}(\bar{\varepsilon}_1 + 0.5\bar{\varepsilon}_2) + \bar{D}\bar{\Theta}_1(\bar{\kappa}_1 + 0.5\bar{\kappa}_2)],$$

$$N_2^o = \delta_3 \delta_1 [\bar{B}(\bar{\varepsilon}_2 + 0.5\bar{\varepsilon}_1) + \bar{D}\bar{\Theta}_1(\bar{\kappa}_2 + 0.5\bar{\kappa}_1)],$$

$$M_1^o = \delta_3^2 \delta_2 \bar{D}[\bar{\Theta}_1(\bar{\varepsilon}_1 + 0.5\bar{\varepsilon}_2) + \bar{\kappa}_1 + 0.5\bar{\kappa}_2],$$

$$M_2^o = \delta_3^2 \delta_1 \bar{D}[\bar{\Theta}_1(\bar{\varepsilon}_2 + 0.5\bar{\varepsilon}_1) + \bar{\kappa}_2 + 0.5\bar{\kappa}_1]. \tag{2.2.56}$$

Expressions for Θ_1 и $\bar{\Theta}_1$ can be represented as follows:

$$\Theta_1 = \frac{2}{3} \cdot \frac{(\eta - 1)}{\varepsilon^2} \kappa_\varepsilon, \quad \bar{\Theta}_1 = \frac{2}{3} \cdot \frac{(\eta - 1)}{\bar{\varepsilon}^2} \bar{\kappa}_\varepsilon, \tag{2.2.57}$$

where κ_ε и $\bar{\kappa}_\varepsilon$ are determined by formulas

$$\kappa_\varepsilon = 2[\varepsilon_1 \kappa_1 + 0.5(\varepsilon_1 \kappa_2 + \varepsilon_2 \kappa_1) + \varepsilon_2 \kappa_2],$$

$$\bar{\kappa}_\varepsilon = 2[\bar{\varepsilon}_1 \bar{\kappa}_1 + 0.5(\bar{\varepsilon}_1 \bar{\kappa}_2 + \bar{\varepsilon}_2 \bar{\kappa}_1) + \bar{\varepsilon}_2 \bar{\kappa}_2]. \tag{2.2.58}$$

Let's move on to dimensionless quantities by formulas:

$$\{\tilde{\alpha}_o, \tilde{r}_o\} = \{\alpha_o, r_o\}/R_*,$$

$$\{\tilde{N}_1^o, \tilde{N}_2^o, \tilde{Q}^o, \tilde{\bar{N}}_1^o, \tilde{\bar{N}}_2^o\} = R_*(E_* h_*^2)^{-1}\{N_1^o, N_2^o, Q^o, \bar{N}_1^o, \bar{N}_2^o\},$$

$$\{\tilde{\bar{T}}^o, \tilde{\bar{\Psi}}^o, \tilde{Y}_1, \tilde{Y}_2\} = (E_* h_*^2)^{-1}\{\bar{T}^o, \bar{\Psi}^o, Y_1, Y_2\},$$

$$\{\tilde{M}_1^o, \tilde{M}_2^o\} = R_*(E_* h_*^3)^{-1}\{M_1^o, M_2^o\},$$

$$\{\tilde{M}^o, \tilde{Y}_3\} = (E_* h_*^3)^{-1}\{M^o, Y_3\},$$

$$\{\tilde{\bar{\kappa}}_1, \tilde{\bar{\kappa}}_2, \tilde{K}_1, \tilde{K}_2\} = R_*\{\bar{\kappa}_1, \bar{\kappa}_2, K_1, K_2\},$$

$$\{\tilde{w}, \tilde{u}, \tilde{Y}_4, \tilde{Y}_5\} = R_*^{-1}\{w, u, Y_4, Y_5\}, \quad \tilde{Y}_6 = Y_6,$$

$$\tilde{p} = R_*^2(E_* h_*^2)^{-1}p, \quad \tilde{h}_o = h_o/h_*,$$

$$\tilde{\sigma} = \sigma/\sigma_l, \tilde{\bar{\varepsilon}} = \bar{\varepsilon}/\bar{\varepsilon}_l, \quad \tilde{\bar{\Theta}}_1 = \bar{\Theta}_1 R_*. \tag{2.2.59}$$

The defining relations (2.2.56) in the dimensionless form are given as

$$\tilde{N}_1^o = \delta_3\delta_2 k_\sigma[(\tilde{\tilde{B}}/\varepsilon_*)(\bar{\varepsilon}_1 + 0.5\bar{\varepsilon}_2) + \varepsilon_*\tilde{\tilde{D}}\tilde{\tilde{\Theta}}_1(\tilde{\tilde{\kappa}}_1 + 0.5\tilde{\tilde{\kappa}}_2)],$$

$$\tilde{N}_2^o = \delta_3\delta_1 k_\sigma[(\tilde{\tilde{B}}/\varepsilon_*)(\bar{\varepsilon}_2 + 0.5\bar{\varepsilon}_1) + \varepsilon_*\tilde{\tilde{D}}\tilde{\tilde{\Theta}}_1(\tilde{\tilde{\kappa}}_2 + 0.5\tilde{\tilde{\kappa}}_1)],$$

$$\tilde{M}_1^o = \delta_3^2\delta_2 k_\sigma\tilde{\tilde{D}}[\tilde{\tilde{\Theta}}_1(\bar{\varepsilon}_1 + 0.5\bar{\varepsilon}_2) + \tilde{\tilde{\kappa}}_1 + 0.5\tilde{\tilde{\kappa}}_2],$$

$$\tilde{M}_2^o = \delta_3^2\delta_1 k_\sigma\tilde{\tilde{D}}[\tilde{\tilde{\Theta}}_1(\bar{\varepsilon}_2 + 0.5\bar{\varepsilon}_1) + \tilde{\tilde{\kappa}}_2 + 0.5\tilde{\tilde{\kappa}}_1]. \qquad (2.2.60)$$

Here,

$$\varepsilon_* = \frac{h_*}{R_*}, \quad k_\sigma = \frac{\sigma_B}{E_*}, \quad \tilde{\tilde{B}} = \frac{4\tilde{\Lambda}(\tilde{\tilde{\varepsilon}})\tilde{h}_o}{3},$$

$$\tilde{\tilde{D}} = \frac{\tilde{\Lambda}(\tilde{\tilde{\varepsilon}})\tilde{h}_o^3}{9}, \quad \tilde{\tilde{\Theta}}_1 = \frac{2}{3} \cdot \frac{(\eta-1)}{\varepsilon_v^2 \cdot \tilde{\tilde{\varepsilon}}^2}\tilde{\tilde{\kappa}}_\varepsilon; \qquad (2.2.61)$$

$$\tilde{\tilde{\kappa}}_\varepsilon = 2[\bar{\varepsilon}_1\tilde{\tilde{\kappa}}_1 + 0.5(\bar{\varepsilon}_1\tilde{\tilde{\kappa}}_2 + \bar{\varepsilon}_2\tilde{\tilde{\kappa}}_1) + \bar{\varepsilon}_2\tilde{\tilde{\kappa}}_2]. \qquad (2.2.62)$$

Next, we will use a variant of the DN, with respect to which there are allegations of experimental confirmation of its operability for metals [31].

Chapter 3

Membranes in Systems for the Protection of Equipment from Destruction by Excessive Pressure

The information related to the tasks of modeling and technologies of manufacture of safety clapping membranes, a high-precision operation used in the protection device against overload pressure, is provided here.

Convex dome shells are used as elements of safety systems devices that protect industrial equipment and tanks against overpressure destruction. At a certain level of pressure acting on their convex part, one can observe loss of stability in these shells followed by destruction that keeps the integrity of the protected equipment. These shells are referred to as flapping safety membranes (FSM) because of the characteristic sound of flap accompanying the action of the stability loss. Other terms include buckling safety membranes (BSM), bursting discs (BD).

Destroyed shell elements (bursting discs), which are a part of membrane safety devices, have resulted in the search for trusty protection frames of the equipment in chemical and a petroleum-refining industry, and, afterwards, in thermal and atomic engineering.

The formation of experimental and theoretical research, development of the FSM designs, and manufacturing technologies at Southern Federal University are closely connected with I.I. Vorovich who was not only the head of Mechanics and Applied Mathematics Research Institute but was also actively involved into research in nonlinear theory of shells' stability.

Research intensity of the FSM is determined by a variety of protective devices designs, structural complexity of membranes' response mechanisms, and their sensitivity to small stochastic deviations from idealized design schemes.

3.1.   Experiments and Manufacturing Techniques

Flapping membranes are one of the most effective destructible elements of the safety devices [45, 56]. There are also other types of membranes, triggered directly from the action of critical hydrostatic pressure: bursting, rupture, shearing, tearing, combined membranes and membranes made for special purposes. Further classification can be also performed on the design features and techniques for determining subcritical and post-critical mechanical behavior: smooth membranes; membrane comprising notches or weakening zones; membranes of complicated geometric shapes (corrugated or with variable curvature of a meridian); composite, layered structurally anisotropic membranes; twisted destructible elements of the shell type; embitter membranes; etc. Membranes are also distinguished by materials, protective coatings, and a type of fixing support, i.e. rigid clamping (flat or conical clamp), elastic, and free support [8].

There are many factors in the technologies for mass production of FSM that significantly affect the difference in the critical loads of real shells and design models such as perfectly spherical clamped elastic domes. These factors are referred to as initial imperfections and include, first of all, geometric imperfections of small bending type and features of a support in the inflection zone near the FSM's dome. However, in the research process, it turned out that this is only part of the real factors affecting critical loads. There are other deviations from the design models.

(1) It is usually assumed that physical and mechanical properties are averaged in the material of structures and can be represented by simple continuum models with fixed values of elasticity modulus, strengths, etc. These data are insufficient for thin-walled

domes made of flat product. There is a noticeable difference of the physical and mechanical properties of the flat product along and across the direction of rolling.

(2) The sheet materials manufactured by the industry are characterized by a very noticeable thickness difference in sheet area. To reduce the influence of this factor, it is often necessary to look for areas with acceptable thickness variance when choosing work pieces.

(3) Real nature of membrane support, when it is clamped by elements of holders so that the required level of tightness is ensured, cannot be normally identified with usual models of "rigid clamping," "plastic hinge," or "free support". The attempts to evaluate the actual conditions for the membranes supports were unsuccessful. At least, it was not possible to connect the estimates of the microstrains of the clamping rings of the holders with the observed changes in the operating pressure of those types of FSM for which the developers hoped to "idealize" the clamping surfaces by introducing gaskets, plastic or crumpled peripheral rings, stress concentrators, etc.

(4) Micro-structural analysis of the near-edge areas of membranes, which can be obtained by plastic molding, showed that the thickness of the membrane material is non-uniform due to micro-particulates; local zones of tightening; and aeration in the inflection regions.

(5) There is also a group of factors related to the technological characteristics of media, which is in contact with surface of safety membranes. Apart from the well-known factors associated with the corrosion of the material, the analysis of the operating capacity of safety membranes shows the phenomenon of mass transfer from the material of holder rings to the dome of membranes. This can be explained by the dissolution of carbon in the surface layers of parts, with its subsequent adsorption on the marginal zones of the domed part of membranes made of low-carbon austenitic steel, which entails a change in its physical and mechanical properties.

So far, experimental methods are the most reliable in predicting the response of the FSM under pressure loading. The traditional method of determining a critical pressure value for flapping membranes is to break down a number of samples in monitoring tests. The accuracy of critical pressure for individual membrane is relatively low ($\pm 5\%$), since the confidence interval of pressure is established for the whole set on the basis of data dispersion for tested samples.

Generally, destruction tests have been employed at the initial stage, when the experimental data on the behavior of membranes were accumulated. There was an extensive set of experiments performed for smooth shells made of different materials by the traditional technology of free plastic drawing. The geometric characteristics (thickness of a sample, diameter of clamping rings, and height of dome), loading path (multiple loading and unloading), and the clamping force of membranes in the rings were varied in these experiments. Critical heights of drawing out were found, even until the break. The influence of the parameters on critical load of membranes as well as of the spread of the pressure response in the set of nominally identical membranes was analyzed. Methods for predicting critical pressures were developed. Results were analyzed with the help of statistics; dispersions and confidence intervals of critical pressure were estimated. Stability loss forms were classified.

Considerable experience, gained during this research, including the study of the membranes behavior in the subcritical stage of deformation, allowed researchers to develop methods and invent hardware support systems for nondestructive tests, to improve the design of membranes and obtain serial samples of high accuracy. The issue of making the FSM with high accuracy response was solved by creation of the equipment for manufacturing membranes and development of the instrumentation to predict the pressure response by non-destructive methods.

Developed methods of non-destructive determination of the pressure response of membranes are based on physically independent principles of diagnosis. In the first group of methods, extrapolation of the critical load on nonlinear diagram "pressure–displacement" is used; in the second group, direct acoustic emission method or employment of acoustically active coatings is used; in the third group, optical methods is used. By analyzing subcritical behavior, membranes were divided into two classes — thin-walled and thick-walled membranes.

Acoustic emission activity for membranes of the first type was observed just before the "flap" in the form of a very short pulse spike of AE; for membranes of the second type, it was in the nature of asymptotic growth. It was concluded that, in the first case, subcritical elastic deformation is occurring, while the second case is characterized with elastic–plastic deformations and progressing extension of plastic deformation zones.

These two types of membranes also different in terms of compression marks on the initial stage of stability loss. For thin membranes, the initial compression marks of stability loss were localized, mainly in the near-edge zone. Rarely, there were cases when initial development of compression marks was observed in the area close to the center of the dome. Thick-walled membranes were characterized by only near-edge compression marks of a crescent (conchoidal) type.

It should be noted that the above classification differences are clearly observed for membranes having a variation in the critical pressure not worse than 1–2%. In addition, for membrane with high accuracy actuation, buckling modes were surprisingly similar. Otherwise, compression marks of ill-defined geometry were observed. If the influence of initial imperfections is considered, then, undoubtedly, the set of membranes with stable value of critical loads was characterized by the same type of deviations from the ideal spherical shell (with similar main parameters). The scatter of pressure values is associated with variations in the initial imperfections. Usually, this is determined by the properties of the flat product used to make dome shells (variation in thickness, non-uniformity, and anisotropy of physical and mechanical properties).

These observations led to the idea of controlling properties of the shell by making artificial initial imperfections, which ensure stable buckling modes, critical load, and are of great importance for material of membranes, aging under real operating conditions.

One such trend, implemented in the construction of membranes and their manufacturing technologies, was aimed to provide initial axisymmetric buckling modes and initialization of clipping from the top of the dome. It was beneficial for stability loss dynamics and efficiency of the section "opening" during slitting of the membrane by cutters. Artification can be constructive when stiffness parameters are controlled, and geometric, for which the form of the middle surface can be changed.

Fig. 3.1.1.   Structurally anisotropic membranes.

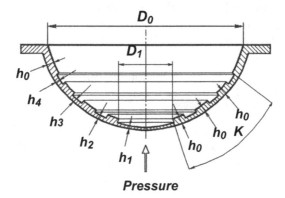

Fig. 3.1.2.   Membrane of step-variable thickness.

One of the proposed options for constructive artification is as follows. The membrane is made stepwise-variable thickness, Figs. 3.1.1 and 3.1.2. The membrane has circular grooves, the depth of which increases from the periphery to the center. The central zone is homogeneous, with the smallest thickness in the vicinity of the dome top. The optimal diameter of the central zone is approximately 0.4 of the membrane diameter. The weakened zone initiates the beginning of loss of stability in the vicinity of the apex, and the higher rigidity of the structural anisotropy in the annular direction contributes to the development of axisymmetric deformation. The manufacturing process of such a shell consists of several stages: applying a photolayer to the plate blank, exposing it to a negative pattern of concentric rings, photodevelopment and fixing, drawing a domed shell, increasing the

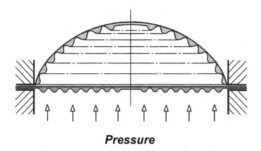

Pressure

Fig. 3.1.3. Corrugated membrane.

resistance of the photolayer, and chemical etching of the grooves. Only chemically active materials, such as aluminum, are suitable for this purpose.

For chemically resistant materials, constructive anisotropy can be created by the corrugation, Figs. 3.1.1a and 3.1.3. An appropriate method of manufacturing was proposed. A ring corrugation is stamped on a flat blank and a self-hardening sealant, which is not prone to adhesion, is applied. A flat smooth plate is placed on the sealant. The resulting "sandwich" is clamped in rings, and the dome is extracted. After that, the corrugated shell is separated from the support shell and released from the sealant.

The structurally artificated shells can also include membranes with rigid overlays in the area of the top of the dome, Fig. 3.1.1 (membrane b) and Fig. 3.1.4. Pads replace external knives and are designed to ensure effective destruction of the membrane during operation.

Further investigations of different types of membranes showed that the smooth thin-walled class of membranes has the best characteristics, with smooth thick-walled membranes in the second place. Membranes with notches, grooves, and other stress concentrators ensuring effective opening have lower operating performance. This is due to the accumulation of micro-plastic damage occurring during repeated loads and reducing cycle-to-cycle pressure response.

That is why further studies were concentrated on the geometric artificiality [23, 24], the purpose of which was the production of smooth membranes with smaller value of curvature in the tip vicinity. This is possible with the help of additional influences to the main

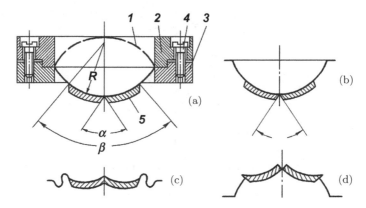

Fig. 3.1.4.   Membrane with rigid overlays.

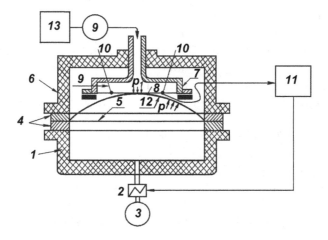

Fig. 3.1.5.   The device with back pressure chamber.

forming pressure. Three methods have been proposed to implement this approach.

In one of the devices, along with the main chamber for forming the shell, there is a back pressure chamber with a flexible plate, Fig. 3.1.5.

While drawing out is happening on the final stage, the dome-shaped shell is in contact with the flexible plate, which causes resistance in apex zone. As a result, smaller curvature can be observed in that zone, in contrast with the curvature obtained by free drawing out. It should be mentioned that sensors controlling the curvature value of the shell as well as backpressure are used.

In the second method, the dome at the final stage of forming came into contact with a layer of elastic material, the stiffness of which was selected experimentally. The third method uses the application of concentrated force at the top of the shell. This method is the most technologically intensive and is currently used as the main method. At the same time, the modern implementation of this manufacturing and forecasting technology is based on the methods of computer control and mathematical processing of information by the software and hardware complex "ASD-Membrane", Fig. 4.4.1. The accuracy of the prediction of the pressure membrane is 1%–3%, in the operating temperature range of 20–450°C. The window of the automated diagnostic system in the process of non-destructive testing of the membrane is shown in Fig. 4.5.1.

From the standpoint of the sensitivity of the shells to the initial random technological imperfections, the artification can be interpreted as an artificial creation of "imperfections" that overlap the influence of random ones, stabilize the critical load, and provide an advantageous form of buckling from the top of the shell. The artification eliminates the problems associated with the boundary zone and eliminates the impact of other random imperfections associated with the quality of the original sheet materials, since the adjustment for a given critical load is made individually for each membrane. The shape of the buckling provides a high dynamics of motion of the shell after buckling and the effective destruction and almost complete disclosure of the cross-section gap in the contour through the use of circular knives.

3.2. Development of Mathematical Modeling

In direct mathematical modeling based on the solution of boundary value problems in the first stages, BSM was considered as a segment of an elastic spherical shell of constant thickness. The problem of nonlinear stability of such a shell loaded with uniform external pressure has long attracted the attention of many researchers. This model stimulated the development of the concept of buckling "in the big," the theory of bifurcations of equilibrium forms, the geometric theory of shell stability, the theory of sensitivity of critical loads to initial imperfections, application of asymptotic methods to shell stability analysis, development and implementation of direct numerical methods, and algorithms for solving nonlinear boundary value problems on a computer. We studied the influence of the complicated mechanical properties of the material; it was considered and backed by a layered shell, and this solved the dynamic problem as well as several other ones.

When comparing the theoretical and experimental results, agreement was observed in single cases of precision experiments, in which measures were taken to comply with idealized conditions, usually assumed in theory. For example, very thin, highly elastic shells obtained by spraying copper in vacuum on a spherical steel substrate were tested. In another version, very flat spherical segments were cut from a thick plate blank. On its contour, the ring through which jamming was carried out was left. Such approaches can be considered as a step of experiment towards proving the theory.

The vast majority of experimental results on critical loads are lower than theoretical ones for elastic shells by several times. The reason, as noted, is the random initial imperfections. Therefore, semi-empirical formulas were usually used for the preliminary selection of parameters in the manufacture of membranes by free drawing pressure.

For coordination of mathematical models with physical ones for real shells, it was necessary to counter-move the theory to the experimental conditions, taking into account their geometric and physical-mechanical properties more adequately.

At first, methods of modeling geometric imperfections developed on the theory of flat elastic shells. The initial deflection was often set close to the expected form of buckling. The boundary conditions were also varied under the assumption of possible compliance of the shell contour sealing. The essential factors of more adequate modeling include the need to take into account the work of the shell material beyond the elastic limit. In the presence of initial imperfections, the role of plasticity can increase significantly, but the joint account of these factors was practically not carried out.

In [11], the geometrical aspects of the dome shape setting, corresponding to the technology of plastic forming of flapping membranes by the method of free plastic drawing by pressure, was considered.

According to this technology, a flat round thin plate — the work-piece — is clamped in the annular flanges so that the edges of the workpiece do not slip during the molding process. The flange is hermetically attached to the container into which the medium (air, oil) is pumped under pressure. Transmitted pressure plastically deforms the workpiece, turning it into a dome. The process of free drawing is stable and allows you to get a fairly high-lift shell. In this method of forming, there is an inflection zone from the flat flange to the dome. This zone of negative Gaussian curvature was modeled by the element of a circular torus smoothly conjugated to the dome part. The main dome was considered as an ellipsoid of rotation. When compared with the equivalent height of the spherical shell, calculations showed significant differences in curvature. In addition, attention was paid to the variability of thickness and the possibility of its simple approximation based on the analysis of experiments [30]. Taking into account these factors, computational modeling was carried out in [13]. It showed the tendency of convergence of the calculated bifurcation and experimental critical loads of flapping membranes, losing stability in asymmetric forms. In comparison with idealized models, critical loads were reduced by 2–3 times. Moreover, the presence of a torus marginal zone (flange) dramatically rearranges critical forms. If for the considered idealized models the number of circumferential waves was from 6 to 18, then for shells with a collar this number was equal to two. This corresponds qualitatively to the form of buckling often observed in experiments, when deformations develop from the boundary zone in the form of a crescent dent.

Significant progress on the path of convergence of theory and experiment was made in [16–20]. To do this, it was necessary to solve the problems of two stages. The first stage simulates the manufacture of an articulated membrane. At the second stage, the stability of the obtained shell is analyzed. The mathematical model of deformation of physically nonlinear shells of rotation at large displacements and angles of rotation taking into account the change of the metric is developed and used for the forming stage. The defining relations of Davis–Nadai type, taking into account the heterogeneity of the material properties induced by deformations in thickness, are derived. Using the semi-inverse method, it was possible to construct approximate analytical solutions to the problem of forming spherical and ellipsoidal domes from plates and theoretically justify the concept of articulation with the use of concentrated force. Comparison of theory and experiment showed their agreement [19].

Flapping membranes made by the new technology have been tested under operating conditions at the stand of the State Scientific Center of the Russian Federation — Physics and Power Engineering Institute named after Academician A.I. Leipunsky. The test results in normal operation conditions confirmed the compliance of the membranes with strict technical standards for accuracy and stability of the actuation pressure, as well as durability, including under the influence of sodium coolant.

Chapter 4

Stretching of Dome Shell
from a Circular Plate

In the manufacturing of domed shells from sheet materials, the simplest and most technologically advanced method is the free drawing by uniform pressure. The sheet blank is clamped between two round flanges and loaded on one side by air or liquid pressure (oil), causing the transformation of the plate into a dome-shaped shell under conditions of large plastic deformations. This method is also used to construct tensile diagrams that determine the properties of the material and to test sheet materials for the deformation limit [30, 32, 39]. There is a similar method of forming shells of rotation of cylindrical blanks.

4.1. The Complexity of Building a Step-by-Step Iterative Algorithm

Let us consider a system of equations for the problem of forming domed shells of round plate blanks. In the case of a plate, the initial angles of inclination of the normal and curvature are equal to zero: $\Phi_o = 0$, $k_1^o = k_2^o = 0$. The differential system of equations for the plate follows from the equilibrium equations and differential kinematic relations:

$$\bar{T}^{o'} = \alpha_o r_o \delta_1 \delta_2 p \cos \Phi,$$

$$\bar{\Psi}^{o'} = \alpha_o \bar{N}_2^o - \alpha_o r_o \delta_1 \delta_2 p \sin \Phi,$$

61

$$M^{o'} = \alpha_o M_2^o \cos \Phi + \alpha_o r_o \delta_1 Q^o,$$

$$w' = \alpha_o \delta_1 \sin \Phi,$$

$$u' = \alpha_o (\delta_1 \cos \Phi - 1),$$

$$\Phi' = \alpha_o K_1. \tag{4.1.1}$$

Here, r_o (polar radius) can be taken as an independent radial coordinate; p^o is the intensity of uniform hydrostatic (tracking) pressure

$$\delta_1 = (1 + \varepsilon_1), \quad \delta_2 = (1 + \varepsilon_2),$$

$$\delta_3 = (1 + \varepsilon_3), \quad K_2 = (\sin \Phi)/r_o,$$

$$\bar{T}^o = r_o \bar{V}^o, \quad \bar{\Psi}^o = r_o \bar{H}^o, \quad M^o = r_o M_1^o,$$

$$\bar{N}_1^o = \bar{V}^o \sin \Phi + \bar{H}^o \cos \Phi,$$

$$Q^o = \bar{V}^o \cos \Phi - \bar{H}^o \sin \Phi,$$

$$\bar{N}_1^o = N_1^o + (K_1 M_1^o + K_2 M_2^o)/\delta_1,$$

$$\bar{N}_2^o = N_2^o + (K_1 M_1^o + K_2 M_2^o)/\delta_2. \tag{4.1.2}$$

In physical relations (2.2.50)–(2.2.52), the values κ_1 and κ_2 are expressed through the curvature of the resulting dome:

$$\kappa_1 = -\delta_3 K_1 = -\frac{K_1}{\delta_1 \delta_2} = -\frac{\bar{K}_1}{\delta_2},$$

$$\kappa_2 = -\delta_3 K_2 = -\frac{K_2}{\delta_1 \delta_2} = -\frac{\bar{K}_2}{\delta_1}. \tag{4.1.3}$$

Then, in a simple version, the defining relations of the Davis–Nadai (DN) type are given the form:

$$N_1^o = \bar{B}_1 (\bar{\varepsilon}_1 + 0.5 \bar{\varepsilon}_2),$$

$$N_2^o = \bar{B}_2 (\bar{\varepsilon}_2 + 0.5 \bar{\varepsilon}_1);$$

$$M_1^o = -\bar{D}_1 (\bar{K}_1 + 0.5 \bar{K}_2),$$

$$M_2^o = -\bar{D}_2 (\bar{K}_2 + 0.5 \bar{K}_1), \tag{4.1.4}$$

where

$$\bar{\varepsilon}_1 = \ln(\delta_1), \quad \bar{\varepsilon}_2 = \ln(\delta_2),$$

$$\bar{\bar{D}}_1 = \bar{D}_1 \delta_3 = \frac{h_o^3 \Lambda(\bar{\varepsilon}) \delta_3^3 \delta_2}{9}, \quad \bar{\bar{D}}_2 = \bar{D}_2 \delta_3 = \frac{h_o^3 \Lambda(\bar{\varepsilon}) \delta_3^3 \delta_1}{9},$$

$$\bar{B} = \frac{4\Lambda(\bar{\varepsilon})h_o}{3}, \quad \bar{D} = \frac{4\Lambda(\bar{\varepsilon})h_o^3}{3},$$

$$\bar{B}_1 = \frac{\bar{B}}{\delta_1} = \frac{4\Lambda(\bar{\varepsilon})h_o}{3\delta_1}, \quad \bar{B}_2 = \frac{\bar{B}}{\delta_2} = \frac{4\Lambda(\bar{\varepsilon})h_o}{3\delta_2}. \qquad (4.1.5)$$

It is logical to take the displacements u, w, the angle of inclination of the normal in the deformed state Φ, and the generalized force components $\bar{T}^o = r_o \bar{V}^o$, $\bar{\Psi}^o = r_o \bar{H}^o$, $M^o = r_o M_1^o$ that are included in the natural boundary conditions (1.4.6) as the main functions. We introduce the unified notation for them as follows:

$$Y_1 = \bar{T}^o, \quad Y_2 = \bar{\Psi}^o, \quad Y_3 = M^o,$$

$$Y_4 = w, \quad Y_5 = u, \quad Y_6 = \Phi. \qquad (4.1.6)$$

In order to apply the method of translating the boundary value problem into Cauchy problems, it is necessary to form a resolving system of differential equations of the boundary value problem on the basis of kinematic relations (4.1.1) and equilibrium equations (4.1.2) containing derivatives of the basic functions. To build the algorithm, it is necessary to bring the system to a normalized form, i.e. the right parts of the system in the canonical form of the record must be expressed through the main functions Y_j, $j = 1, \ldots, 6$. For this purpose, the remaining kinematic and determining relations (4.1.1) and (4.1.2) are involved.

Due to the complex nonlinearity, this cannot be done explicitly. Therefore, in one of the constructed algorithms, the method of continuation by the load parameter in combination with the iterative process was used.

Let us form a system of relatively dimensionless quantities, the transition to which is performed in accordance with the formulas (2.2.60). We have the following relations:

$$\tilde{\kappa}_1 = -\delta_3 \tilde{K}_1, \quad \tilde{\kappa}_2 = -\delta_3 \tilde{K}_2,$$

$$\tilde{\tilde{\kappa}}_1 = -\delta_3 \tilde{\tilde{K}}_1, \quad \tilde{\tilde{K}}_1 = \tilde{K}_1/\delta_1, \quad \tilde{\tilde{\kappa}}_2 = -\delta_3 \tilde{\tilde{K}}_2, \quad \tilde{\tilde{K}}_2 = \tilde{K}_2/\delta_2; \quad (4.1.7)$$

$$\varepsilon_* = \frac{h_*}{R_*}, \quad k_\sigma = \frac{\sigma_l}{E_*}, \quad \tilde{\tilde{B}} = \frac{4\tilde{\Lambda}(\tilde{\tilde{\varepsilon}})\tilde{h}_o}{3}, \quad \tilde{\tilde{D}} = \frac{\tilde{\Lambda}(\tilde{\tilde{\varepsilon}})\tilde{h}_o^3}{9}. \qquad (4.1.8)$$

The diagram of material properties is transferred to the unit plane:

$$\tilde{\sigma}(\tilde{\bar{\varepsilon}}) = \tilde{C}_l \tilde{\bar{\varepsilon}}^\eta, \quad \tilde{C}_l = \tilde{C} \tilde{\bar{\varepsilon}}_l^\eta,$$

$$\tilde{\Lambda}(\tilde{\bar{\varepsilon}}) = \tilde{C}_\Lambda \tilde{\bar{\varepsilon}}^{(\eta-1)}, \quad \tilde{C}_\Lambda = \tilde{C} \tilde{\bar{\varepsilon}}_l^{(\eta-1)}. \tag{4.1.9}$$

In a simple version, the defining relations of the DN type have the form:

$$\tilde{N}_1^o = (k_\sigma/\varepsilon_*)\tilde{\bar{B}}_1(\bar{\varepsilon}_1 + 0.5\bar{\varepsilon}_2),$$

$$\tilde{N}_2^o = (k_\sigma/\varepsilon_*)\tilde{\bar{B}}_2(\bar{\varepsilon}_2 + 0.5\bar{\varepsilon}_1),$$

$$\tilde{M}_1^o = -k_\sigma \tilde{\bar{\bar{D}}}_1(\tilde{\bar{K}}_1 + 0.5\tilde{\bar{K}}_2),$$

$$\tilde{M}_2^o = -k_\sigma \tilde{\bar{\bar{D}}}_2(\tilde{\bar{K}}_2 + 0.5\tilde{\bar{K}}_1), \tag{4.1.10}$$

where

$$\bar{\varepsilon}_j = \ln(1 + \varepsilon_j) = \ln \delta_j, \quad j = 1, 2, 3;$$

$$\tilde{\bar{B}}_1 = \tilde{\bar{B}}/\delta_1, \quad \tilde{\bar{B}}_2 = \tilde{\bar{B}}/\delta_2,$$

$$\tilde{\bar{D}}_1 = \delta_3^3 \delta_2 \tilde{\bar{D}}, \quad \tilde{\bar{D}}_2 = \delta_3^3 \delta_1 \tilde{\bar{D}}. \tag{4.1.11}$$

In the variant of the defining relations for the plate following from (2.2.61), we obtain

$$\tilde{N}_1^o = \delta_3\delta_2 k_\sigma[(\tilde{\bar{B}}/\varepsilon_*)(\bar{\varepsilon}_1 + 0.5\bar{\varepsilon}_2) - \varepsilon_*\delta_3\tilde{\bar{D}}\tilde{\bar{\Theta}}_1(\tilde{\bar{K}}_1 + 0.5\tilde{\bar{K}}_2)],$$

$$\tilde{N}_2^o = \delta_3\delta_2 k_\sigma[(\tilde{\bar{B}}/\varepsilon_*)(\bar{\varepsilon}_2 + 0.5\bar{\varepsilon}_1) - \varepsilon_*\delta_3\tilde{\bar{D}}\tilde{\bar{\Theta}}_1(\tilde{\bar{K}}_2 + 0.5\tilde{\bar{K}}_1)],$$

$$\tilde{M}_1^o = \delta_3^2\delta_2 k_\sigma \tilde{\bar{D}}[\tilde{\bar{\Theta}}_1(\bar{\varepsilon}_1 + 0.5\bar{\varepsilon}_2) - \delta_3(\tilde{\bar{K}}_1 + 0.5\tilde{\bar{K}}_2)],$$

$$\tilde{M}_2^o = \delta_3^2\delta_2 k_\sigma \tilde{\bar{D}}[\tilde{\bar{\Theta}}_1(\bar{\varepsilon}_1 + 0.5\bar{\varepsilon}_2) - \delta_3(\tilde{\bar{K}}_2 + 0.5\tilde{\bar{K}}_1)]; \tag{4.1.12}$$

$$\tilde{\bar{\Theta}}_1 = \frac{2}{3} \cdot \frac{(\eta - 1)}{\varepsilon_v^2 \cdot \bar{\bar{\varepsilon}}^2}\tilde{\bar{\kappa}}_\varepsilon,$$

$$\tilde{\bar{\kappa}}_\varepsilon = -2\delta_3[\bar{\varepsilon}_1\tilde{\bar{K}}_1 + 0.5(\bar{\varepsilon}_1\tilde{\bar{K}}_2 + \bar{\varepsilon}_2\tilde{\bar{K}}_1) + \bar{\varepsilon}_2\tilde{\bar{K}}_2]. \tag{4.1.13}$$

On the basis of (4.1.2) and the relations (4.1.10), we have

$$\tilde{N}_1^o = \frac{k_\sigma}{\varepsilon_*} \tilde{B}_1(\bar{\varepsilon}_1 + 0.5\bar{\varepsilon}_2) + \varepsilon_* \left(\frac{\tilde{M}^o}{\tilde{r}_o} \frac{\tilde{K}_1}{\delta_1} + \tilde{M}_2^o \frac{\tilde{K}_2}{\delta_1} \right),$$

$$\tilde{N}_2^o = \frac{k_\sigma}{\varepsilon_*} \tilde{B}_2(\bar{\varepsilon}_2 + 0.5\bar{\varepsilon}_1) + \varepsilon_* \left(\frac{\tilde{M}^o}{\tilde{r}_o} \frac{\tilde{K}_1}{\delta_2} + \tilde{M}_2^o \frac{\tilde{K}_2}{\delta_2} \right); \qquad (4.1.14)$$

$$\tilde{K}_1 = \frac{\tilde{M}^o}{k_\sigma \tilde{r}_o \tilde{D}_1} - \frac{\delta_1}{2\delta_2} \tilde{K}_2; \quad \tilde{M}_2^o = -k_\sigma \tilde{D}_2 \left(\tilde{K}_2 + \frac{\delta_2}{2\delta_1} \tilde{K}_1 \right). \quad (4.1.15)$$

After the introduction of the dimensionless vector of the main functions, we obtain:

$$\tilde{Y}_1 = \tilde{T}^o, \ \ \tilde{Y}_2 = \tilde{\Psi}^o, \ \ \tilde{Y}_3 = \tilde{M}^o, \ \ \tilde{Y}_4 = \tilde{w}, \ \ \tilde{Y}_5 = \tilde{u}, \ \ \tilde{Y}_6 = \Phi. \quad (4.1.16)$$

The differential equation (4.1.1) are of the following form:

$$\tilde{Y}_1' = \tilde{\alpha}_o \tilde{r}_o \delta_1 \delta_2 \tilde{p} \cos \tilde{Y}_6,$$

$$\tilde{Y}_2' = \tilde{\alpha}_o \tilde{N}_2^o - \tilde{\alpha}_o \tilde{r}_o \delta_1 \delta_2 \sin \tilde{Y}_6,$$

$$\tilde{Y}_3' = \tilde{\alpha}_o \tilde{M}_2^o \cos \tilde{Y}_6 + \varepsilon_*^{-1} \tilde{\alpha}_o \tilde{r}_o \delta_1 \tilde{Q}^o,$$

$$\tilde{Y}_4' = \tilde{\alpha}_o \delta_1 \sin \tilde{Y}_6,$$

$$\tilde{Y}_5' = \tilde{\alpha}_o (\delta_1 \cos \tilde{Y}_6 - 1),$$

$$\tilde{Y}_6' = \tilde{\alpha}_o \tilde{K}_1. \qquad (4.1.17)$$

We denote further $\tilde{r}_o = \xi$ and consider R_* equal to r_p (the radius of the plate), so $\tilde{\alpha} = 1$. Through the entered functions, basic ones are easily determined as follows:

$$\bar{\varepsilon}_2 = \frac{\tilde{Y}_5}{\xi}, \quad \tilde{K}_2 = \frac{\sin \tilde{Y}_6}{\xi}, \quad \tilde{N}_1^o = \frac{1}{\xi}(\tilde{Y}_1 \sin \tilde{Y}_6 + \tilde{Y}_2 \cos \tilde{Y}_6). \quad (4.1.18)$$

Further, we obtain from the relations (4.1.17) after the transformations

$$\tilde{N}_1^o = \frac{k_\sigma}{\varepsilon_*} \tilde{B}_1(\bar{\varepsilon}_1 + 0.5\bar{\varepsilon}_2) - \frac{\varepsilon_*}{\delta_1 \xi^2}[(\tilde{Y}_3)^2/(k_\sigma \tilde{D}_1) + 0.75 k_\sigma \tilde{D}_2 (\sin \tilde{Y}_6)^2],$$

$$\tilde{N}_2^o = \frac{k_\sigma}{\varepsilon_*} \tilde{B}_2(\bar{\varepsilon}_2 + 0.5\bar{\varepsilon}_1) - \frac{\varepsilon_*}{\delta_2 \xi^2}$$

$$\times [(\tilde{Y}_3)^2/(k_\sigma \tilde{D}_1) + 0.75 k_\sigma \tilde{D}_2 (\sin \tilde{Y}_6)^2]; \quad (4.1.19)$$

$$\tilde{K}_1 = \frac{-\tilde{Y}_3}{k_\sigma \tilde{D}_1 \xi} - \frac{\delta_1}{2\delta_2} \tilde{K}_2 = \frac{-\tilde{Y}_3}{k_\sigma \tilde{D}_1 \xi} - \frac{\delta_1}{2\delta_2} \cdot \frac{\sin \tilde{Y}_6}{\xi},$$

$$\tilde{M}_2^o = -0.75 k_\sigma \tilde{D}_2 \frac{\sin \tilde{Y}_6}{\xi} + \frac{\delta_1}{2\delta_2} \cdot \frac{\tilde{Y}_3}{\xi}. \quad (4.1.20)$$

Substituting in the first of the relations (4.1.20) the expression $\tilde{N}_1^o$ through the basic functions, we obtain the equation to determine $\bar{\varepsilon}_1$.

$$\frac{1}{\xi}(\tilde{Y}_1 \sin \tilde{Y}_6 + \tilde{Y}_2 \cos \tilde{Y}_6)_1 = \frac{k_\sigma}{\varepsilon_*} \tilde{B}_1(\bar{\varepsilon}_1 + 0.5\bar{\varepsilon}_2)$$

$$- \frac{\varepsilon_*}{\delta_1 \xi^2}\left[\frac{(\tilde{Y}_3)^2}{k_\sigma \tilde{D}_1} + 0.75 k_\sigma \tilde{D}_2 (\sin \tilde{Y}_6)^2\right]_1. \quad (4.1.21)$$

This equation is essentially nonlinear, so it is not possible to solve it explicitly. But it can be considered linear $\bar{\varepsilon}_1$ with respect to the metric state of the shell corresponding to the previous step of the iterative process. After determination of $\bar{\varepsilon}_1$, the values $\tilde{N}_2^o$, $\tilde{M}_2^o$, $\tilde{K}_1$, included in the right-hand side of the system (4.1.17), are expressed in terms of the main functions $\tilde{Y}_j$ of the second ratio (4.1.19) and formulas (4.1.20).

After performing substitutions in equations (4.1.17) and transformations, we obtain the following system:

$$\tilde{Y}_1' = \xi(1 + \varepsilon_1)(1 + \tilde{Y}_5/\xi)\tilde{p} \cos \tilde{Y}_6,$$

$$\tilde{Y}_2' = \frac{k_\sigma}{\varepsilon_*} \tilde{B}_2(\ln(1 + \tilde{Y}_5/\xi) + 0.5 \ln(1 + \varepsilon_1)) + \frac{\varepsilon_*}{\xi^2(1 + \tilde{Y}_5/\xi)}$$

$$\times \left[\frac{(\tilde{Y}_3)^2}{k_\sigma \tilde{D}_1} - \frac{(1+\varepsilon_1)}{(1+\tilde{Y}_5/\xi)} \cdot \tilde{Y}_3 \sin \tilde{Y}_6 - 0.75 k_\sigma \tilde{D}_2 (\sin \tilde{Y}_6)^2 \right]$$

$$- \xi(1+\varepsilon_1)(1+\tilde{Y}_5/\xi)\tilde{p} \sin \tilde{Y}_6,$$

$$\tilde{Y}_3' = \left(\frac{(1+\varepsilon_1)}{2(1+\tilde{Y}_5/\xi)} \cdot \frac{\tilde{Y}_3}{\xi} - 0.75 k_\sigma \tilde{D}_2 \frac{\sin \tilde{Y}_6}{\xi} \right) \cos \tilde{Y}_6$$

$$+ \varepsilon_*^{-1}(1+\varepsilon_1)(\tilde{Y}_1 \cos \tilde{Y}_6 - \tilde{Y}_2 \sin \tilde{Y}_6),$$

$$\tilde{Y}_4' = (1+\varepsilon_1)\sin \tilde{Y}_6, \quad \tilde{Y}_5' = (1+\varepsilon_1)\cos \tilde{Y}_6 - 1,$$

$$\tilde{Y}_6' = \frac{\tilde{Y}_3}{k_\sigma \xi} \cdot \frac{1}{\tilde{D}_1} - \frac{(1+\varepsilon_1)}{2\xi(1+\tilde{Y}_5/\xi)} \sin \tilde{Y}_6. \tag{4.1.22}$$

Here, $1 + \varepsilon_1 = \delta_1(\xi, \tilde{Y}_1, \tilde{Y}_2, \tilde{Y}_3, \tilde{Y}_5, \tilde{Y}_6)$ is a function from the variables in parentheses.

Attempts have been made to implement the algorithm based on the method of continuation by the load parameter in combination with the iterative process. In the first iteration, the material properties and the shell metric for a given load level are taken from the previous level, which is a small step away.

One of the difficulties in the implementation of the algorithm is to select the initial approximation of the parameters of the Cauchy problems, which is reduced to the boundary value problem of the applied method of two-sided zeroing. This was overcome by the solution of the problem of low load with the use of private variants of a system in an elastic area of shallow shells.

Another aspect of the model and algorithm is to set the boundary conditions on the support loop. During plastic deformation, a sharp bend occurs at the edge of the plate with the formation of a zone such as a plastic hinge. One of the easiest ways of modeling is to set kinematic conditions as fixed hinge. More adequate is the setting of the connection of the edge moment and the angle of rotation. However, in order to identify the coupling factor, appropriate experiments should be developed and carried out.

As for the removal of the coordinate singularity at the vertex, the corresponding conditions can be formulated on the basis of the homogeneity of the stress–strain state in the vicinity of the pole. Another easier way is to "cut out" the singularity and set homogeneous boundary conditions either on the contour of the small hole or on the boundary of the rigid washer in the vicinity of the pole.

Work with this algorithm revealed significant difficulties in time of the account and numerical stability. Therefore, efforts have been made towards the construction of approximate analytic solutions through the use of integrated capabilities of MathCad. Highly effective solutions were obtained.

4.2. Semi-inverse Method in the Task of Spherical Dome Forming

Initially, an analytical solution to the problem of plastic drawing of a spherical dome from a round plate was constructed. In this case, the conditions under which the molded dome retains sphericity were clarified, and the technique of constructing the solution was practiced.

The solution is constructed by a semi-inverse method based on the above mathematical model. We recall that the mathematical model takes into account large displacements and rotation angles, changes in the metric, and compression of the material in thickness. The power approximation of the diagram of the hardening material and deformation of physical relations with logarithmic elongations are used. It is assumed that the relative elongation e_k, $\varepsilon_k = e_k|_{\zeta=0}$, $k = 1, 2, 3$ is comparable to one; compression of the material normal $e_3 = \varepsilon_3$ is constant in thickness. The used kinematic relations have the form

$$e_1 = (\varepsilon_1 + \zeta\kappa_1), \quad \kappa_1 = \Phi'_o/\alpha_o - \delta_3 K_1,$$

$$\varepsilon_1 = (w' \sin \Phi + u' \cos \Phi)/\alpha_o + \cos(\Phi - \Phi_o) - 1,$$

$$e_2 = (\varepsilon_2 + \zeta\kappa_2), \quad \varepsilon_2 = u/r_o,$$

$$\kappa_2 = (\sin \Phi_o)/r_o - \delta_3 K_2, \quad \gamma = \gamma_o/(1 + \varepsilon_1),$$

$$K_1 = \Phi'/\alpha_o, \quad K_2 = (\sin \Phi)/r_o,$$

$$\gamma_o = (w' \cos \Phi - u' \sin \Phi)/\alpha_o - \sin(\Phi - \Phi_o). \qquad (4.2.1)$$

Here, Φ_o and Φ are the angles of inclination of the material normal to the axis of rotation before and after deformation; κ_1 and κ_2 are the characteristics of the change of major curvatures; $\delta_k = 1 + \varepsilon_k$, $k = 1, 2, 3$. For the plate in the initial state, the angle of inclination of the normal to the axis of symmetry $\Phi_o = 0$. The angle of the transverse shear γ is assumed to be small and then reset to zero.

The record of a normalized system of differential equations in a dimensionless form for a plate loaded with uniform pressure has the form

$$d\tilde{\tilde{T}}^o/d\xi = \tilde{\alpha}_o \xi \delta_1 \delta_2 \tilde{p} \cos \Phi,$$

$$d\tilde{\tilde{\Psi}}^o/d\xi = \tilde{\alpha}_o \tilde{N}_2^o - \tilde{\alpha}_o \xi \delta_1 \delta_2 \tilde{p} \sin \Phi,$$

$$d\tilde{M}^o/d\xi = \tilde{\alpha}_o \tilde{M}_2^o \cos \Phi + \tilde{\alpha}_o \xi \delta_1 \tilde{Q}^o / \varepsilon_*,$$

$$d\tilde{w}/d\xi = \tilde{\alpha}_o \delta_1 \sin \Phi,$$

$$d\tilde{u}/d\xi = \tilde{\alpha}_o (\delta_1 \cos \Phi - 1),$$

$$d\Phi/d\xi = \tilde{\alpha}_o \tilde{K}_1, \tag{4.2.2}$$

where

$$\tilde{\tilde{T}}^o = \xi \tilde{V}^o, \quad \tilde{\tilde{\Psi}}^o = \xi \tilde{H}^o, \quad \tilde{M}^o = \xi \tilde{M}_1^o;$$

$$\tilde{V}^o = \tilde{N}_1^o \sin \Phi + \tilde{Q}^o \cos \Phi, \quad \tilde{H}^o = \tilde{N}_1^o \cos \Phi - \tilde{Q}^o \sin \Phi,$$

$$\tilde{N}_1^o = \tilde{V}^o \sin \Phi + \tilde{H}^o \cos \Phi, \quad \tilde{Q}^o = \tilde{V}^o \cos \Phi - \tilde{H}^o \sin \Phi; \tag{4.2.3}$$

$$\tilde{N}_1^o = \tilde{N}_1^o + \varepsilon_* (\tilde{K}_1 \tilde{M}_1^o + \tilde{K}_2 \tilde{M}_2^o)/\delta_1,$$

$$\tilde{N}_2^o = \tilde{N}_2^o + \varepsilon_* (\tilde{K}_1 \tilde{M}_1^o + \tilde{K}_2 \tilde{M}_2^o)/\delta_2. \tag{4.2.4}$$

When normalizing, it was supposed $h_* = h_p$, $R_* = r_p$, so that an independent radial variable $\xi = \tilde{r}_o \in [0, 1]$, the Lame coefficient $\tilde{\alpha}_o = 1$. Here, $\varepsilon_* = h_*/R_* = h_p/r_p$ is the parameter of thinness, $\tilde{p}$ is the intensity of the hydrostatic pressure, $\tilde{V}^o$ and $\tilde{H}^o$ are internal forces oriented along the axis of symmetry and the radius of the cylindrical coordinate system, and $\tilde{M}_1^o$ is a bending moment along the meridian.

It is known that at large plastic deformations, many materials (metals and their alloys) behave practically as incompressible. Therefore, we assume that the relation $\delta_1 \delta_2 \delta_3 = 1$ holds. The appearance of generalized forces (4.2.4) in the derivation of equilibrium equations based on the Lagrange principle is a consequence of taking into account the compression of the material normal and the incompressibility hypothesis.

The material properties are characterized by a loading diagram that is approximated by a power function. The material properties diagram is translated into a unit plane in the coordinates of dimensionless stresses $\tilde{\sigma} = \sigma/\sigma_l$ and relative true deformations $\tilde{\bar{e}} = \bar{e}/\bar{e}_l$, (2.1.10). We use the physical ratio of the Davis–Nadai (DN) for an incompressible material, linking stress and the logarithmic strain in the principal axes (Chapter 2).

In a simplified version of the two-dimensional defining relations of the type DN secant module $\tilde{\Lambda}(\tilde{e}) \approx \tilde{\Lambda}(\tilde{\bar{e}})$, the variability of material properties over thickness is neglected. This is justified in the problems of strong drawing, which are considered. Then the relations of forces and moments with the components of deformations in a dimensionless form obtain the relatively simple form of

$$\tilde{N}_1^o = (k_\sigma/\varepsilon_*)\tilde{\bar{B}}_1(\bar{\varepsilon}_1 + 0.5\bar{\varepsilon}_2),$$

$$\tilde{N}_2^o = (k_\sigma/\varepsilon_*)\tilde{\bar{B}}_2(\bar{\varepsilon}_2 + 0.5\bar{\varepsilon}_1), \qquad (4.2.5)$$

$$\tilde{M}_1^o = -k_\sigma\tilde{\bar{D}}_1(\tilde{\bar{K}}_1 + 0.5\tilde{\bar{K}}_2),$$

$$\tilde{M}_2^o = -k_\sigma\tilde{\bar{D}}_2(\tilde{\bar{K}}_2 + 0.5\tilde{\bar{K}}_1), \qquad (4.2.6)$$

where

$$\bar{\varepsilon}_j = \ln(1 + \varepsilon_j), \quad \tilde{\bar{K}}_j = \tilde{K}_j/\delta_j, \quad j = 1, 2; \quad k_\sigma = \sigma_B/E_*;$$

$$\tilde{\bar{B}}_1 = \tilde{\bar{B}}/\delta_1, \quad \tilde{\bar{B}}_2 = \tilde{\bar{B}}/\delta_2, \quad \tilde{\bar{D}}_1 = \delta_3^3\delta_2\tilde{\bar{D}}, \quad \tilde{\bar{D}}_2 = \delta_3^3\delta_1\tilde{\bar{D}},$$

$$\tilde{\bar{B}} = (4/3)\tilde{\Lambda}(\tilde{\bar{\varepsilon}})\tilde{h}_o, \quad \tilde{\bar{D}} = (1/9)\tilde{\Lambda}(\tilde{\bar{\varepsilon}})\tilde{h}_o^3. \qquad (4.2.7)$$

In the transition to dimensionless quantities as characteristic normalizing values, $h_* = h_p$, $R_* = r_p$, $E_* = \sigma_l$ are accepted.

Let us consider the pressure extraction of a flat round plate with dimensionless thickness $\tilde{h}_p = 1$, pinched along the contour of the radius $\tilde{r}_p$, with a spherical dome (segment) of a given height $\tilde{w}_o$. The curvature of the segment is

$$\tilde{K} = \tilde{R}^{-1} = 2\tilde{w}_o/(\tilde{w}_o^2 + \tilde{r}_p^2), \quad \sin\Phi_c = \tilde{r}_p/\tilde{R}, \quad \cos\Phi_c = 1 - \tilde{w}_o/\tilde{R},$$

where Φ_c is half the angle of the arc solution of the meridian of the dome. Naturally, the main curvature of the shells are equal. $\tilde{K}_1/\delta_1 = \tilde{K}_2/\delta_2 = \tilde{K} = \tilde{R}^{-1}$. It follows that $\delta_2 = (\tilde{R}/\xi)\sin\Phi$.

When drawing in the vicinity of the ground loop, a sharp bend appears, and it is in this zone that the flat pinched part of the plate passes into the dome. As experiments show, this zone is very narrow (about half the thickness of the plate) so that its size can be neglected and taken for the dome $\tilde{r}_p = 1$. Therefore, this zone is conditionally cut off, and its force influence on the circuit is replaced by reactions — vertical and horizontal linear forces, as well as the moment.

Based on the analysis of physical experiments on the molding process [30] and comparative computational experiments, it was found that the variable thickness of the dome is well approximated by the quadratic dependence

$$\tilde{h}(\Phi) = \tilde{h}_p\{1 - \delta[1 - (\Phi/\Phi_c)^2]\}, \tag{4.2.8}$$

where $\delta = 1 - \tilde{h}_o/\tilde{h}_p = 1 - h_o/h_p$ is a relative thinning of the dome at the top. Then,

$$\varepsilon_3 = -\delta[1 - (\Phi/\Phi_c)^2]. \tag{4.2.9}$$

Due to the incompressibility of the material during deformation, the volume of the plate and the resulting dome remain unchanged. Therefore, the equation can be written from the condition of equal volumes of the plate and the dome to determine the coefficient δ

$$2\pi\tilde{R} \int_0^{\Phi_2} \tilde{h}(\Phi)\tilde{R}\sin\Phi d\Phi = \pi\tilde{r}_p^2\tilde{h}_p, \tag{4.2.10}$$

from which

$$\delta = \frac{\Phi_c^2[1 - \cos\Phi_c - 0.5(\tilde{r}_p/\tilde{R})^2]}{2(1 + 0.5\Phi_c^2 - \Phi_c\sin\Phi_c - \cos\Phi_c)}. \tag{4.2.11}$$

We now apply the incompressibility condition to the parts of the plate and shell cut off by parallels with the current Lagrangian coordinate ξ

$$2\pi\tilde{R} \int_0^{\Phi(\xi)} \tilde{h}(\Phi)\tilde{R}\sin\Phi d\Phi = \pi\xi^2\tilde{h}_p. \tag{4.2.12}$$

Integration (4.2.12) gives the following functional equation to determine the function $S(\xi) \equiv \sin(\Phi(\xi))$:

$$S(\xi) = \frac{[(\arcsin(S(\xi)) - \Phi_c^2 - 2)\delta + \Phi_c^2]\sqrt{1 - [S(\xi)]^2} - f(\xi)}{2\delta \cdot \arcsin(S(\xi))}, \tag{4.2.13}$$

where

$$f(\xi) = (1 - \delta)\Phi_c^2 - 2\delta - 0.5(\Phi_c\xi/\tilde{R})^2. \qquad (4.2.14)$$

The equation (4.2.13) is effectively solved by the iterative method based on the principle of compressive maps [26]. As an initial approximation for $S(\xi)$, you can take the function

$$S_0(\xi) = \tilde{R}/[\xi + \tilde{u}_0(\xi)], \qquad (4.2.15)$$

where one of the options $\tilde{u}_0(\xi)$ can be taken as

$$\tilde{u}_0(\xi) = \tilde{u}_{a0}\xi(1 - \xi). \qquad (4.2.16)$$

Another option is even better

$$\tilde{u}_0(\xi) = \tilde{u}_{a0}\tilde{R}\Phi(\xi)(1 - \Phi(\xi)/\Phi_c). \qquad (4.2.17)$$

The coefficient $\tilde{u}_{a0}$ is determined by the application at the top of the dome of the incompressibility and homogeneity of the deformed state of the middle surface $\varepsilon_1(0) = \varepsilon_2(0)$, which gives

$$\tilde{u}_{a0} = \frac{1}{\sqrt{1 - \delta}} - 1. \qquad (4.2.18)$$

As shown by the calculations, successive approximations of the iterative process quickly converge at the second iteration. After exiting the process, the values are determined

$$\Phi(\xi) = \arcsin(S(\xi)), \quad \cos\Phi(\xi) = \{1 - [(S(\xi)]^2\}^{1/2}, \quad (4.2.19)$$

$$\delta_2(\xi) = \tilde{R} \cdot S(\xi)/\xi, \quad \varepsilon_2(\xi) = \delta_2(\xi) - 1,$$

$$\delta_1(\xi) = [\delta_2(\xi)\delta_3(\xi)]^{-1}, \quad \varepsilon_1(\xi) = \delta_1(\xi) - 1, \qquad (4.2.20)$$

as well as internal efforts (4.2.5) and moments (4.2.6).

Then the system (4.2.2) is integrated. We introduce notations for the following integrals:

$$I_{11}(\xi) = \int_0^\xi [\xi/\delta_3(\xi)] \cdot \cos\Phi(\xi)d\xi;$$

$$I_{22}(\xi) = \int_0^\xi [\xi/\delta_3(\xi)] \cdot \sin\Phi(\xi)d\xi,$$

$$I_{41}(\xi) = \int_0^\xi \delta_1(\xi) \cdot \sin \Phi(\xi) d\xi,$$

$$I_{51}(\xi) = \int_0^\xi \delta_1(\xi) \cdot \cos \Phi(\xi) d\xi,$$

$$I_{21}(\xi) = \int_0^\xi \tilde{N}_2^o(\xi) d\xi. \tag{4.2.21}$$

From the first, second, fourth, and fifth equations of the system (4.2.2) it follows that

$$\tilde{T}^o(\xi) = \tilde{p} \cdot I_{11}(\xi) + \tilde{P}_0/(2\pi),$$

$$\tilde{\Psi}^o = I_{21}(\xi) - \tilde{p} \cdot I_{22}(\xi) + r_p \tilde{H}_c; \tag{4.2.22}$$

$$\tilde{w}(\xi) = -\tilde{w}_o + I_{41}(\xi), \quad \tilde{u}(\xi) = I_{51}(\xi). \tag{4.2.23}$$

The displacement $\tilde{w}(\xi)$ vanishes on the edge contour at $\xi = 1$, and $\tilde{u}(1)$ close to zero. The values $\tilde{P}_0$ and $\tilde{H}_c$ are the integration constants. In this case, $\tilde{P}_0$ means the force concentrated at the apex and $\tilde{H}_c$ is a linear power radial direction on the contour.

On the basis of integrals (4.2.23), ratios (4.2.3), and the third equation of system (4.2.2), it is possible to define pressure of forming of a dome of the set height $\tilde{w}_o$

$$\tilde{p} = \frac{[I_{21}(\xi) + \tilde{r}_p \tilde{H}_c] \sin \Phi(\xi) - [\tilde{P}_0/(2\pi)] \cos \Phi(\xi) + \varepsilon_* \tilde{\tau}_1(\xi)/\delta_1(\xi)}{I_{11}(\xi) \cos \Phi(\xi) + I_{22}(\xi) \sin \Phi(\xi)} \tag{4.2.24}$$

where

$$\tilde{\tau}_1(\xi) = \tilde{M}_1^o(\xi)\{1 - [\delta_1(\xi)/\delta_2(\xi)] \cos \Phi(\xi)\} + \xi \cdot [d\tilde{M}_1^o(\xi)/d\xi]. \tag{4.2.25}$$

Since the pressure is uniform, the expression (4.2.24) must be constant. In order to bring it as close as possible to the constant, it is necessary to set non-zero values $\tilde{P}_0$ and $\tilde{H}_c$. It follows that the spherical shape of the dome is provided not only by hydrostatic pressure and clamping reactions of the boundary contour but also by the force concentrated at the top. At the same time, computational experiments show that in the process of drawing the shell, the force must act in the direction of applying pressure and decrease to

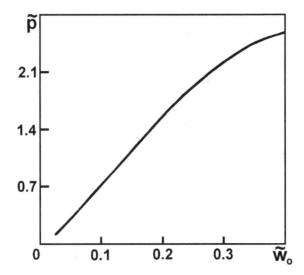

Fig. 4.2.1. Pressure–dome height dependence.

zero at a certain height of the dome. After this height, the force concentrated at the top should increase and act in the opposite direction.

In Figs. 4.2.1 and 4.2.2, the dependences are shown in a dimensionless form, respectively, of the forming pressure and the force at the apex on the relative height of the dome $\tilde{w}_o$, [21]. The ratio of the thickness of the initial plate to its radius $\varepsilon_* = h_p/r_p$ was set to 0.0038. The diagram of the properties of the plate material corresponds to steel 12X18H10T.

During the forming process, the pressure and force must vary from zero to the corresponding dome height values in proportion to time. The sign of the force applied at the apex occurs at the relative height of the dome $\tilde{w}_o \approx 0.2$.

Thus, with a strictly spherical molding of the shell segment, the force at the apex, acting together with the drawing pressure, is required. It follows that when the dome is freely drawn by pressure, the shell passes through stages in which the deviation from the spherical shape first occurs in the direction of the flattened ellipsoid, and then in the direction of the elongated ellipsoid of rotation. This confirms the conclusions of works [11, 39]. Therefore, a more general solution should be sought in the class of spheroids.

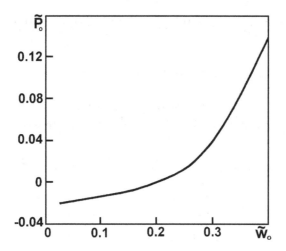

Fig. 4.2.2. Connection of power at the top with dome height.

4.3. Forming a Spheroidal Shell with a Small Eccentricity

Let us consider the pressure extraction from a flat circular plate with a thickness h_p, clamped along the contour of the radius r_p, spheroidal dome (segment) of a given height w_o, [16] The meridian of the segment corresponds to the arc of an ellipse with half-axes a_e and b_e, the square of eccentricity $e_x^2 = 1 - (b_e/a_e)^2$, which is significantly less than one. The equation of an ellipse has the parametric representation: $x = a_e \sin t$, $z = b_e \cos t$. Here because of the smallness of the differences between the parameter t identified with Φ, the angle between the normal to the meridian of the dome and the axis of symmetry at the boundary of the segment is denoted Φ_c. In the transition to dimensionless quantities, as in the previous problem, we accept $h_* = h_p$, $R_* = r_p$, $E_* = \sigma_l$. Then, dimensionless geometric parameters are $\tilde{h}_p = 1$, $\tilde{r}_p = 1$, $\tilde{a}_e = a_e/r_p$, $\tilde{b}_e = b_e/r_p$, $e_x^2 = 1 - (\tilde{b}_e/\tilde{a}_e)^2$. The explanations for the inflection zone remain the same as in Section 4.2.

As in the previous section, we assume that the variable thickness of the dome in the dimensionless form is adequately determined by the quadratic dependence of the form

$$\tilde{h}(\Phi) = \tilde{h}_p\{1 - \delta[1 - (\Phi/\Phi_c)^2]\},$$

where $\delta = 1 - \tilde{h}_0/\tilde{h}_p$ is a relative thinning of the dome at the top. Then,

$$\varepsilon_3 = -\delta[1 - (\Phi/\Phi_c)^2].$$

The value δ is determined from the condition of equality of the volume of the plate and the ellipsoidal shell with a small eccentricity

$$2\pi a_e \int_0^{\Phi_c} \tilde{h}(\Phi) a_e [1 - 0.5 e_x^2 (\sin \Phi)^2] \sin \Phi d\Phi = \pi \tilde{r}_p^2 \tilde{h}_p. \quad (4.3.1)$$

Hence,

$$\delta = [I_0 - 0.5(\tilde{r}_p/\tilde{a}_e)^2]/I_1, \quad (4.3.2)$$

where

$$I_0 = C_c\{(e_x^2/6)[(S_\Gamma)^2 + 2] - 1\} + 1 - (1/3)e_x^2,$$
$$I_1 = I_{11} + I_{12},$$
$$I_{11} = [3\Phi_c S_c^3 + C_c S_c^2 + 20(C_c - 1)$$
$$+ 18\Phi_c S_c - 9\Phi_c^2](e_x/\Phi_c)^2/27,$$
$$I_{12} = [2 + \Phi_c^2 - 2(C_c + \Phi_c S_c)]/\Phi_c^2,$$
$$C_c = \cos\Phi_c, \quad S_c = \sin\Phi_c. \quad (4.3.3)$$

The incompressibility condition is applicable to the parts of the plate and shell cut off by parallels with the current Lagrangian coordinate ξ. The condition of equality of such volumes in the considered case has the form

$$2\pi\tilde{a}_e \int_0^{\Phi(\xi)} \tilde{h}(\chi) a_e [1 - 0.5 e_x^2 (\sin\chi)^2](\sin\chi) d\chi = \pi\xi^2 \tilde{h}_p. \quad (4.3.4)$$

After integration, this relation is reduced to a functional equation to determine the radial displacement $\tilde{u}(\xi)$

$$\tilde{u}(\xi) = \tilde{a}_e\{S_3[S_\Phi(\xi)]^3 + S_2 C_\Phi(\xi)[S_\Phi(\xi)]^2$$
$$+ S_0 C_\Phi(\xi) - F_1(\xi)\} - \xi, \quad (4.3.5)$$

where

$$\tilde{x}(\xi) = \xi + \tilde{u}(\xi), \quad \Phi(\xi) = \arcsin(S_\Phi(\xi)),$$

$$S_\Phi(\xi) = \sin(\Phi(\xi)) = \tilde{x}(\xi)/\tilde{a}_e,$$

$$C_\Phi(\xi) = \cos(\Phi(\xi)) = [1 - (S_\Phi(\xi))^2]^{1/2}; \quad (4.3.6)$$

$$f_1(\xi) = \xi^2/(2\tilde{a}_e^2),$$

$$f_2(\xi) = f_1(\xi) - (1 - \delta)[1 - (1/3)e_x^2] + (2\delta/\Phi_c^2)[1 - (10/27)e_x^2],$$

$$F_1(\xi) = -3\Phi_c^2 f_2(\xi)/[2\delta \cdot \Phi(\xi)(3 - e_x^2)],$$

$$S_1 = 6(3 - e_x^2)^{-1}, \quad S_3 = e_x^2 S_1,$$

$$S_0(\xi) = \frac{\Phi(\xi)/2 - S_1\{[9(1 - \delta)\Phi_c^2 - 20\delta]e_x^2 + 27[(\delta - 1)\Phi_c^2 + 2\delta]\}}{3\delta \cdot \Phi(\xi)},$$

$$S_2(\xi) = \frac{-S_3\{(3/2)\Phi(\xi) + [9(1 - \delta)]\Phi_c^2 - 2\delta]\}}{6\delta \cdot \Phi(\xi)}. \quad (4.3.7)$$

The right parts of equation (4.3.5) depend on $\tilde{u}(\xi)$ through formulas (4.3.6), (4.3.7). Equation (4.3.5) is reduced to the form prepared for the application of the iterative method based on the principle of compressive maps [26]. Functions (4.2.16) and (4.2.17) can be taken as an initial approximation. The process converges in two iterations. Further, after the process is completed, the values (4.3.6) and strain components are determined

$$\varepsilon_2(\xi) = \tilde{u}(\xi)/\xi, \quad \delta_2(\xi) = 1 + \varepsilon_2, \quad \delta_1(\xi) = [\delta_2(\xi)\delta_3(\xi)]^{-1},$$

$$\varepsilon_1(\xi) = \delta_1(\xi) - 1,$$

as well as internal forces and moments (4.2.5), (4.2.6).

Then the system (4.2.2) is integrated. The values of the first, second, fourth and fifth equations are determined as

$$\tilde{T}^o(\xi) = \tilde{p} \cdot I_{11}(\xi) + \tilde{P}_0/(2\pi),$$

$$\tilde{\Psi}^o = I_{21}(\xi) - \tilde{p} \cdot I_{22}(\xi) + r_p \tilde{H}_c; \quad (4.3.8)$$

$$\tilde{w}(\xi) = -\tilde{w}_o + I_{41}(\xi), \quad \tilde{u}(\xi) = I_{51}(\xi), \quad (4.3.9)$$

where

$$I_{11}(\xi) = \int_0^{\xi} [\zeta/\delta_3(\zeta)] \cdot \cos \Phi(\zeta) d\zeta,$$

$$I_{22}(\xi) = \int_0^{\xi} [\zeta/\delta_3(\zeta)] \cdot \sin \Phi(\zeta) d\zeta,$$

$$I_{21}(\xi) = \int_0^{\xi} \tilde{N}_2^o(\zeta) d\zeta, \quad I_{41}(\xi) = \int_0^{\xi} \delta_1(\zeta) \cdot \sin \Phi(\zeta) d\zeta,$$

$$I_{51}(\xi) = \int_0^{\xi} \delta_1(\zeta) \cdot \cos \Phi(\zeta) d\zeta. \tag{4.3.10}$$

The displacement $\tilde{w}(\xi)$ vanishes on the boundary contour (at $\xi = 1$), and $\tilde{u}(\xi)$ is close to zero in the same place. The values $\tilde{P}_0$ and $\tilde{H}_c$ are the integration constants. In this case, $\tilde{P}_0$, it makes sense to concentrate at the top of the force that controls the certification. $\tilde{H}_c$ is the linear force of the radial direction on the contour, and it has the meaning of the reaction of the boundary fixation.

On the basis of (4.3.6), taking into account the kinematic and physical relations from the third equation of the system (4.2.2), the forming pressure of the ellipsoidal dome of a given height $\tilde{w}_o$ is determined. This can be done in two ways — differential, as in the previous section, and integral. Since the geometry of the obtained shell is parametrically known, it is possible to determine the moments and their derivatives through its curvatures and obtained expressions of relative elongations. This is the first approach that gives

$$\tilde{p} = \frac{[I_{21}(\xi) + \tilde{r}_p \tilde{H}_c] \sin \Phi(\xi) - [\tilde{P}_0/(2\pi)] \cos \Phi(\xi) + \varepsilon_* \tilde{\tau}_1(\xi)/\delta_1(\xi)}{I_{11}(\xi) \cos \Phi(\xi) + I_{22}(\xi) \sin \Phi(\xi)}$$

$$\approx const, \tag{4.3.11}$$

where

$$\tilde{\tau}_1(\xi) = \tilde{M}_1^o(\xi)\{1 - [\delta_1(\xi)/\delta_2(\xi)] \cos \Phi(\xi)\} + \xi \cdot [d\tilde{M}_1^o(\xi)/d\xi]. \tag{4.3.12}$$

The output of the expression (4.3.11) for a constant is controlled by parameters $\tilde{P}_0$, $\tilde{H}_c$, e_x.

In the second method, the third equation is integrated, which makes it possible to introduce another integration constant that has the meaning of the bending moment on the contour:

$$\tilde{p} = \frac{\xi \tilde{M}_1^o(\xi) + \tilde{r}_p \tilde{M}_c - J_{31}(\xi) - J_{33}(\xi) + J_{34}(\xi) + J_{36}(\xi)}{J_{32}(\xi) + J_{35}(\xi)}, \approx const$$

(4.3.13)

where

$$J_{31}(\xi) = \int_0^\xi \tilde{M}_2^o \cos \Phi(\zeta) d\zeta,$$

$$J_{32}(\xi) = \frac{1}{\varepsilon_*} \int_0^\xi \delta_1(\zeta) I_{11}(\zeta) \cos \Phi(\zeta) d\zeta,$$

$$J_{33}(\xi) = \frac{\tilde{P}_0}{2\pi\varepsilon_*} \int_0^\xi \delta_1(\zeta) \cos \Phi(\zeta) d\zeta,$$

$$J_{34}(\xi) = \frac{1}{\varepsilon_*} \int_0^\xi \delta_1(\zeta) I_{21}(\zeta) \sin \Phi(\zeta) d\zeta,$$

$$J_{35}(\xi) = \frac{1}{\varepsilon_*} \int_0^\xi \delta_1(\zeta) I_{22}(\zeta) \sin \Phi(\zeta) d\zeta,$$

$$J_{36}(\xi) = \frac{\tilde{H}_c}{\varepsilon_*} \int_0^\xi \delta_1(\zeta) \sin \Phi(\zeta) d\zeta.$$

(4.3.14)

Here, $\tilde{M}_c$ is the third integration constant (edge bending moment).

This method is methodically somewhat better and gives a slightly smoother function, $\tilde{p}$, close to a constant. However, the calculation of the formula (4.3.13) requires more time because it is associated with a double calculation of integrals. A significant acceleration of the account is achieved here using approximations for integrals (4.3.10), (4.3.14). In general, both methods lead to almost the same results.

In addition to the thickness approximation in the form (4.2.10), the following variant was also tested for the ellipsoid class:

$$\tilde{h}(\Phi) = \tilde{h}_p\{1 - \delta(1 - \Phi/\Phi_c) + g \cdot e_x^2[(1 - \Phi/\Phi_c)]^2\}, \qquad (4.3.15)$$

The approximation includes linear and quadratic software components Φ, contains an additional control constant g, and depends on the square of eccentricity. The resulting formulas are more cumbersome.

At small e_x^2, this and previous approximations lead to very close results on the forming pressure. Approximation (4.3.15) can be useful in modeling the forming of more essentially elliptical shells. It is also somewhat more consistent with the distribution of the thickness obtained by the measurements on the real artificated shells.

4.4. Comparison of Theory and Experiment

Physical testing of membranes was performed at the technological installation "ASD-Membrane," developed in the Department of Thin-Walled Structures of the I.I. Vorovich Research Institute of Mechanics and Applied Mathematics of Southern Federal University. The installation allows both the production of buckling safety membranes (BSM) and the prediction of their actuation pressures by non-destructive methods as well as the carrying out of destructive tests.

The automated system "ASD-Membrane" is a hardware–software complex. The complex is designed to implement the following functions: (1) automation of scientific experiments to study the influence of the main mechanical, structural, and technological parameters of the manufacturing process of BSM on the stability of their physical and mechanical properties; (2) automation of the manufacturing process, testing, and non-destructive prediction of the operating pressure of BSM. As is known, BSM are the main destructible elements of safety membrane devices (SMD) included in the emergency protection systems of installations operating under pressure, such as in aggressive, toxic, flammable, and explosive environments.

The system implements the method of determining the opening pressure of the BSM on the characteristic "load–move" without the destruction of samples of membranes. Bringing to destruction can be performed during control tests. The non-destructive method provides high accuracy for determining the critical pressure of each individual membrane. The formation of the operating characteristics of the membranes during their production by the method of artification ensures the accuracy of the response pressure not worse than 1% at room temperatures and not worse than 1–3%, taking into account the operating conditions of operation of the membranes at elevated temperatures, the working interval of which is usually in the range 250–400°C.

The scheme of the installation for the manufacture and prediction of the actuation pressure of BSM is presented in Fig. 4.4.1. Clapping

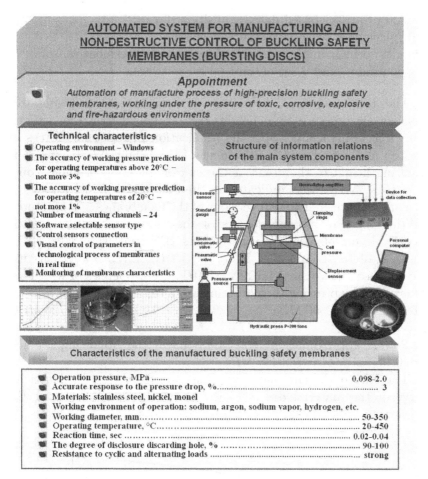

AUTOMATED SYSTEM FOR MANUFACTURING AND NON-DESTRUCTIVE CONTROL OF BUCKLING SAFETY MEMBRANES (BURSTING DISCS)

Appointment

Automation of manufacture process of high-precision buckling safety membranes, working under the pressure of toxic, corrosive, explosive and fire-hazardous environments

Technical characteristics

- Operating environment – Windows
- The accuracy of working pressure prediction for operating temperatures above 20°C – not more 3%
- The accuracy of working pressure prediction for operating temperatures of 20°C – not more 1%
- Number of measuring channels – 24
- Software selectable sensor type
- Control sensors connection
- Visual control of parameters in technological process of membranes in real time
- Monitoring of membranes characteristics

Structure of information relations of the main system components

Characteristics of the manufactured buckling safety membranes

- Operation pressure, MPa 0.098-2.0
- Accurate response to the pressure drop, %.. 3
- Materials: stainless steel, nickel, monel
- Working environment of operation: sodium, argon, sodium vapor, hydrogen, etc.
- Working diameter, mm.. 50-350
- Operating temperature, °C.. 20-450
- Reaction time, sec 0.02-0.04
- The degree of disclosure discarding hole, % 90-100
- Resistance to cyclic and alternating loads .... strong

Fig. 4.4.1. The scheme of installation for manufacturing and forecasting the set pressure of BSM.

membranes of different sizes can be constructed and explored safely on the installation. For each type of BSM, there are appropriate clamping rings of the holder, ensuring the stability of the conditions of the edge fastening of the membrane.

Technically, the system includes the following components:

– power plant for the manufacture and testing of membranes (press, pressure source, electromagnetic pneumatic valves, pressure gauges, a set of clamping rings for different sizes of membranes);

– microprocessor-based data acquisition device;
– displacement sensors, pressure sensor;
– multi-normalizing amplifier of signals of displacement sensors;
– a computer with a set of application software.

The technology of fabrication of membranes has a step-free exhaust air pressure to a height H_1. At height H_1, the artification process associated with the application of resistance forces at the top and acting on the increment of the height ΔH begins. The resulting height of the dome at the end of the hood becomes equal $H = H_1 + \Delta H$ to that indicated in the theoretical model w_o.

For comparison, two real versions of the membranes manufactured for the safety membrane devices of the protection systems of the second circuit of fast neutron reactors (FN-600) were taken. The material of the membranes is stainless steel 12H18H9. Let us use the data for a close material 12X18H10T, the properties of which are determined by pressing the plate through a round hole [32]: $E = 0.21 \cdot 10^6\,MPa$, $\sigma_{0.2} = 360\,MPa$, $\bar{e}_{0.2} = \sigma_{0.2}/E = 0.001714$, $\sigma_l = 720\,MPa$, $\varepsilon_l = 0.615$.

Power approximation constants have the following values: $\eta = 0.1178256$, $C = 762.445\,MPa$.

The form of dimensionless functions approximating the plasticity curve of the material $\tilde{\sigma}(\tilde{e})$ and the secant modulus $\tilde{E}_s(\tilde{e})$ for the considered option is shown in Figs. 4.4.2 and 4.4.3.

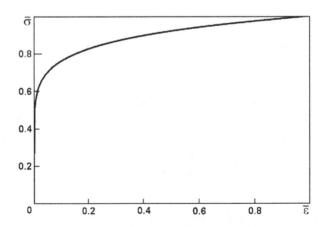

Fig. 4.4.2.   Diagram of the material.

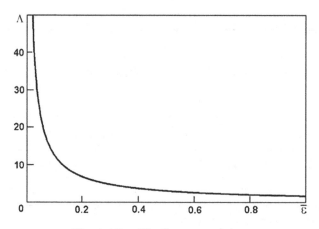

Fig. 4.4.3. The Secant modulus.

Here, $\tilde{C} = 1.059$, $\tilde{C}_l = \tilde{C}\bar{e}_l^{\eta}$ is always equal to one, because $\tilde{C}_l = \tilde{C}\bar{e}_l^{\eta} = C\bar{e}_l^{\eta}/\sigma_l$, $C = \sigma_l/\bar{e}_l^{\eta}$.

Then, $\tilde{C}_E = \tilde{C}\bar{e}_l^{(\eta-1)} = \tilde{C}\bar{e}_l^{\eta}/\bar{e}_l = \tilde{C}\bar{e}_l^{\eta}/\bar{e}_l = 1/\bar{e}_l = 1.626$.

In option 1, the thickness of the workpiece (plate) is equal to 0.28–0.3 mm. The radius of the plate on the ground loop annular flanges is 100 mm. Nominal lifting height is 35–36 mm. The pressure of the hood (molding) $p_f = 1.59$ MPa (or 15.6 ATI). Level of artificial force $P_0 = 6.87$ N (or 0.7 kgf). The nominal critical value of the shell $p_{cr} = 0.255$ MPa (or 2.5 ATI).

In option 2, the thickness of the workpiece (plate) is equal to 0.38–0.4 mm. The radius of the plate on the ground loop annular flanges is 100 mm. Nominal lifting height is 35–36 mm. The pressure of the hood (molding) $p_f = 2.4$ MPa (or 23.5 ATI). Level of artificial force $P_0 = 10.1$ N (or 1.03 kgf). The nominal critical value of the shell $p_{cr} = 0.432$ MPa (or 4.4 ATI).

Tables 4.4.1 and 4.4.2 provide measurements of the thickness distribution (in mm) in the meridian arc for options 1 and 2, respectively. Numbering points from the top.

In Fig. 4.4.4a comparison of the thickness functions (4.2.8) normalized by the thickness of the blank for the shell Type 2, depending on the angle of inclination of the normal, reduced to the interval of unit length is given: $\tilde{\Phi} = \Phi/\Phi_c$. The measurements were made by micrometer with a division value of 0.01 mm, so the relative error is

Table 4.4.1.   Thickness distribution from the top to the contour.

1	2	3	4	5	6	7	8	9	10	11
0.25	0.25	0.25	0.25	0.25	0.26	0.27	0.27	0.29	0.29	0.30

Note: Measurements for option 1.

Table 4.4.2.   Thickness distribution from the top to the contour.

1	2	3	4	5	6	7	8	9	10	11
0.31	0.31	0.31	0.32	0.33	0.33	0.34	0.34	0.35	0.37	0.38

Note: Measurements for option 2.

not less than 3%. The effect on the scatter of experimental points is also exerted by the imperfection of the initial sheet blank. However, a quite adequate agreement between the theoretical and experimental results characterizing the distribution of the relative change in thickness upon drawing is seen. A similar comparison of the variant (4.3.15) is given in Fig. 4.4.4b.

When specifying the same levels of articulating loads in the calculations as in the experiments, the molding pressure according to the formulas (4.3.11), (4.3.13) agrees with the experimental accuracy, 2–3%. The geometry of the segment corresponds to a flattened spheroid with eccentricity $e_x = 0.12$ and the ratio of the half axes

$$k_e = (1 - e_x^2)^{0.5} = 0.993.$$

Dimensionless values that take integration constants for the shell of option 2 are $\tilde{P}_0 = 0.096$, $\tilde{H}_c = 0.9$, $\tilde{M}_c = 0.032$.

Thus, when the shell is extracted from the plate, the technological articulating effect leads to a slightly flattened spheroid with semi-axes differing within one percent.

Let us consider some results of calculations on the example of the shell of option 2. Figure 4.4.5 shows that forming pressure (4.3.13) is close to the constant. Figure 4.4.6 shows curves 1 and 2 of the initial approximation (4.3.5) in variants (4.2.16) and (4.2.17), respectively. Approximations were used to solve the functional equation (4.3.5) in determining the radial displacement $\tilde{u}(\xi)$. Curve 3 corresponds to the first approximation and curve 4 to the second approximation, which

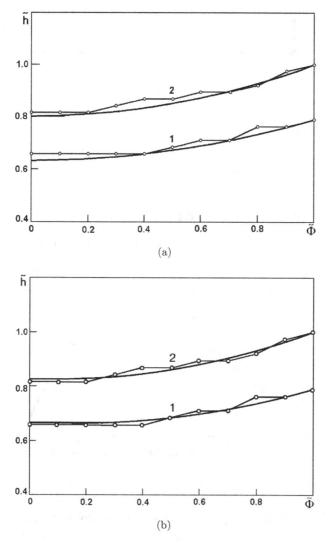

(a)

(b)

Fig. 4.4.4. Comparison of functions (4.2.8) and (4.3.15) — graphs (a) and (b) respectively approximating the thickness of the certified dome. Measurements on two shells.

practically coincides with the exact solution, which demonstrates the fast convergence of the iterative process.

Figure 4.4.7 shows the distributions of the three components of the relative elongations over the radius as a function of the normalized

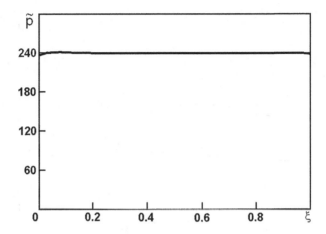

Fig. 4.4.5.   Molding pressure is almost a constant function.

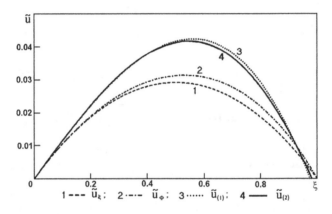

$$1 \; \text{---} \; \tilde{u}_\xi; \quad 2 \; \text{-·-·-} \; \tilde{u}_\phi; \quad 3 \; \text{······} \; \tilde{u}_{(1)}; \quad 4 \; \text{———} \; \tilde{u}_{(2)}$$

Fig. 4.4.6.   Initial approximations and iterations for radial movement.

Lagrangian coordinate ξ. In the area of the dome to $\xi = 0.8$, tangential deformations ε_1 and ε_2 do not differ much, i.e. the deformed state is close enough to uniform. In the edge zone, elongations ε_1 and ε_2 markedly diverge.

The highest elongation values are reached at the top of the dome and are 12% for ε_1 and ε_2, and 20% for ε_3 at a given relative elevation $\tilde{w}_o = 0.35$. The value ε_3 characterizes the thinning of the shell during extraction. There is significant variability in the thickness of the resulting shell from the initial value of the thickness ε_3 of the plate blank to the top $0.8h_p$ of the dome.

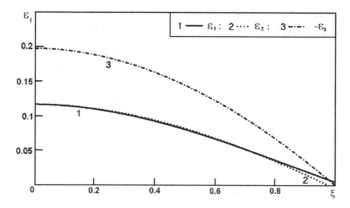

Fig. 4.4.7. The distribution of the relative elongation along the radius.

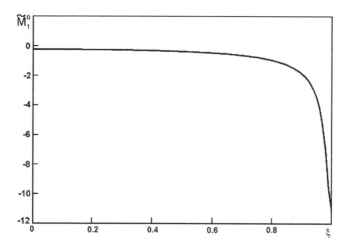

Fig. 4.4.8. Dependence of the bending moment radius.

In Fig. 4.4.8 dependences $\tilde{M}_1^o(\xi)$ and $\tilde{M}_2^o(\xi)$, which correspond to simple constitutive relations (2.2.10), are shown. There is a significant edge effect. In the central part to $\xi = 0.6$, the state close to it is close to a zero-moment and sharply changes in the boundary zone. The boundary effect can be explained not only by the moment reaction of the edge plastic hinge, but also by the variability of the properties of the material changing from the elastic state directly on the clamped circuit, where $\varepsilon_i \approx 0$, to the state with a developed strain intensity on which the secant module depends.

The presence of a pronounced boundary effect is the cause of the numerical instability of the Cauchy problem, which is reduced to the boundary value problem in the algorithm of the method of adjustment. Therefore, the solution of this essentially nonlinear problem by a direct numerical iterative algorithm turns out to be problematic.

The involvement of refined defining relations of the type (2.2.12) gives the results presented in Fig. 4.4.9.

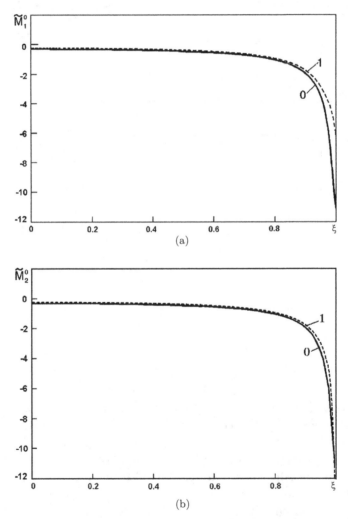

Fig. 4.4.9. The comparison of bending moments for the simple version (0, solid curve) and refined (1, dashed) defining relations: (a) moment 1; (b) moment 2.

Here, the appearance of curves in the vicinity of the boundary contour changes somewhat. The influence of refinement of the defining relations in the considered variants is insignificant on the other results. Thus, the use of simplified defining relations of the type given in (4.2.10) is quite sufficient in this problem.

Variants of the models were compared when the logarithm of elongations in the strain intensity and the defining relations were and were not taken into account. A comparison of these models on the deformations shows that the difference reaches 6% in the zone of the maximum (the top of the dome).

In the theoretical solution, the geometric parameters of the shell and the forming force effects (the pressure level and the articulating force) are set at once. In the real physical process of forming, they change in time. Since the theoretical model uses the power approximation, the loading should be close to the simple one (the theorem of A.A. Ilyushin on the simple loading). That is, the increase in pressure and concentrated force at the top should be proportional to time. Pressure can be accepted for local time. Structurally, it is easier to change the artificating force linearly depending on the height of the dome. This does not exactly correspond to the proportional loading condition. However, the nonlinear dependence "forming pressure–dome height" observed in the experiment [30] does not deviate too much from the linear one.

4.5. Identification of the Chart of Sheet Material Properties

In relation to the problems of strength analysis of thin-walled welded vessels operating under high internal pressure, the technology of testing of sheet materials is developed in [32]. It provides a systematic presentation of the strength of thin-walled vessels used in modern aircraft and other transport installations. Attention is also paid to the generalization of experimental data, as well as the formulation of design and technological recommendations for the design and manufacture of welded thin-walled vessels. In the presented test method and measurement technique, it is required to obtain information from the strain gauges, to control the change in the thickness of the plate during the extraction process, to measure the curvature of the shell at the top.

For problems of superplastic forming in [9], the scheme of the solution of inverse problems of identification of the defining relations by results of technological experiments of molding of shells of simple geometry (cylindrical, spherical) in matrices of the corresponding form on the basis of application of the non-moment theory of shells is offered. Testing is performed either on production equipment or on special installations.

Based on the theoretical and experimental results presented in Sections 4.1–4.4, we can offer a simple way to identify the properties of sheet materials. Using the theoretical solution and some measurements in the process of plastic dome extraction, it is possible to select the parameters of the corresponding approximating curve.

The modern technology of forming dome-shaped flapping membranes includes the stage of free plastic drawing of the plate by hydrostatic pressure up to approximately 85% of the specified height. The further process of forming the dome with increasing pressure is carried out with the effect of the certifying force of a certain value.

The "pressure-lifting height" chart $p_e(H)$ of the first stage can be used to identify the sheet metal hardening chart. A similar problem of mathematical modeling is solved for this purpose (Section 4.1). Based on the comparison of theoretical $p_t(H)$ and experimental $p_e(H)$ diagrams, the coordinates of the reference points of the power-law theoretical relationship $\sigma(\bar{\varepsilon})$ "stress–strain intensity" are adjusted so that $p_t(H)$ and $p_e(H)$ are as close as possible. Moreover, at first it is possible to achieve the coincidence of points with the highest elevation. As shown by numerical experiments, this method already gives results with sufficient accuracy, since the heterogeneity of the sheet properties gives ordinate variations $p_e(H)$ in repeated experiments up to $\pm 4 \div 5\%$. The digitized data $p_e(H)$ for seven sample experiments out of about 60 are presented in Table 4.5.1. The data were obtained on the automated technological installation "ASD-Membrane" presented in Section 4.4. Parameters of circular plates-billets were as follows: diameter $D = 200$ mm, thickness of the billet $h_o = 0.3$ mm, and test temperature $T^oC = 20^o$. The final height of the dome $H \geq 35$ mm. A height of 30 mm is sufficient to identify properties. The material is stainless steel 12X18H9, with a Young's module $E = 2.1 \cdot 10^5$ MPa. For manufacturing and thus for testing stainless steel type 08X18H9, 12X18H10T, and steel 10X13, 20X13, 30X13, monel-metal, brass L-62, L-68, and aluminum can also be used.

Table 4.5.1. The hood pressure $p_{e1}(H)$ and the height of the dome, issued by the system "ASD-Membrane."

H	p_{e1}	p_{e2}	p_{e3}	p_{e4}	p_{e5}	p_{e6}	p_{e7}
0.0	0.0	0.0	0.0	0.0	0.0	0.0	0.0
2.0	0.01	0.01	0.01	0.01	0.01	0.01	0.01
3.5	0.06	0.06	0.05	0.07	0.07	0.06	0.05
6.0	0.15	0.15	0.15	0.17	0.16	0.16	0.15
9.0	0.24	0.24	0.24	0.25	0.25	0.25	0.25
12.0	0.35	0.35	0.34	0.36	0.35	0.35	0.35
15.0	0.46	0.46	0.46	0.48	0.47	0.47	0.47
18.0	0.58	0.58	0.57	0.59	0.59	0.59	0.58
21.0	0.71	0.71	0.71	0.73	0.73	0.72	0.72
24.0	0.85	0.85	0.85	0.87	0.87	0.86	0.86
27.0	1.01	1.00	1.00	1.02	1.01	1.01	1.00
30.0	1.16	1.15	1.16	1.18	1.17	1.17	1.16
33.0	1.32	1.32	1.32	1.34	1.34	1.33	1.32

The data in Table 4.5.1 are removed from the graphs produced by the "ASD-Membrane" system with an accuracy sufficient to reflect the type of curves' "pressure–lifting height." When digitizing by hovering the cursor over the curve, the system provides information about the hit point of measurement and its coordinates with the above accuracy. For example, in Fig. 4.5.1 a variant of the drawing the curve corresponding to the column p_{e1} of the Table 4.5.1 is shown.

More extensive information about the measurements taken (thousands of points) contains database files collected by the automated information retrieval system. After thinning this array, the information is passed to the integrated MathCad package for processing, where the text file is converted to a matrix. On the basis of the matrix, you can build graphs and approximating curves.

Thus, on the basis of the technology of plastic molding and the theoretical solution of the problem by the method of plastic drawing of the dome-shaped shell from a round plate clamped on the outer contour, it is possible to implement a method of identification of the plastic diagram of the material.

The method is tested for metals and makes it possible to simplify the test significantly. There is no need for strain measurement since it is enough to have only the dependencies linking the forming

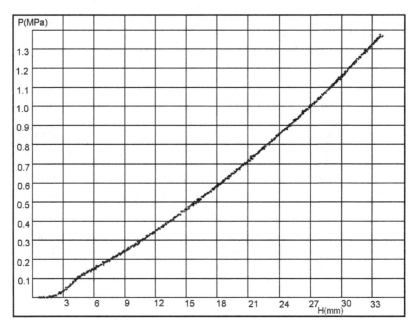

Fig. 4.5.1.   Version of the drawing curve corresponding to column p_{e1} of the Table 4.5.1.

pressure with the movement of the dome top. These relationships are compared with the theoretical model (Section 4.1), which defines all deformation characteristics. In this model, the moment theory for large nonlinear deformations is applied. By minimizing the deviations of the experimental and theoretical loading curves, the parameters of approximation of the material properties diagram are selected, Fig. 4.5.2.

Moreover, since the intensity of deformation along the meridian of the resulting dome varies from zero on the clamping circuit to a certain maximum value at the apex, it is sufficient to compare and lead to a coincidence of the experimental and theoretical height of the shell lifting. Therefore, it is sufficient to measure only the height of the segment and the corresponding exhaust pressure in tests. Then, the mathematical modeling based on the semi-analytical method of solution construction is performed. By varying the values of the material parameter σ_l (sigma limit), it is possible to achieve the coincidence of the theoretical and experimental heights of the dome at the same pressure. The value σ_l that this provides is the

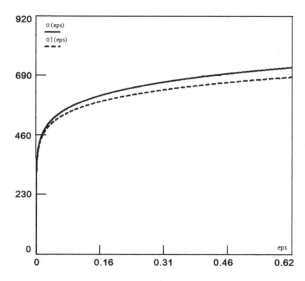

Fig. 4.5.2.   Curves for correcting material properties.

refined parameter of the power approximation of the curve of plastic properties of the material.

Such identification of material properties was tested in the problem of forming of the articulated shells and allowed to improve the coordination of theory and experiment from 3% to 1.5%.

Chapter 5

Solving Problems of Changing the Form of Some Shells of Rotation

5.1. Transformation of Circular Plate into Spheroidal Dome with Significant Eccentricity of Shape

In Chapter 4, the problem of drawing the dome from a round plate, pinched along the boundary contour, was solved under the assumption of a small eccentricity of the resulting spheroid. The realization of this condition was provided by the action of the main forming pressure in combination with the force applied in the apex. In the absence of force at a sufficiently high drawing height, the dome has the shape of an elongated spheroid, the smallness of the eccentricity of which is no longer assumed. A generalization of the semi-inverse method with respect to these conditions is presented here [53].

The construction of the solution is based on the model equations given in Section 4.2. However, a clear capture of integrals of elliptic type, based on the smallness of the eccentricity, is not satisfied.

To localize references to formulas within the section, we provide the necessary information. We recall that the model is based on equations that allow large displacements and angles of rotation, change of the metric of the middle surface, and compression of the material normal. We use power approximations of diagrams of the hardening material and deformation physical relations with logarithmic elongations.

It is believed that the relative elongations e_k, $\varepsilon_k = e_k|_{\zeta=0}$, $k = 1$, $2, 3$, are comparable to the unit; the compression of the material of

95

the normal $e_3 = \varepsilon_3$ is constant in thickness. Kinematic relations have the form:

$$e_1 = (\varepsilon_1 + \zeta\kappa_1), \quad e_2 = (\varepsilon_2 + \zeta\kappa_2),$$

$$\varepsilon_1 = (w'\sin\Phi + u'\cos\Phi)/\alpha_o + \cos(\Phi - \Phi_o) - 1,$$

$$\varepsilon_2 = u/r_o, \quad \kappa_1 = \Phi'_o/\alpha_o - \delta_3 K_1,$$

$$\kappa_2 = (\sin\Phi_o)/r_o - \delta_3 K_2,$$

$$K_1 = \Phi'/\alpha_o, \quad K_2 = (\sin\Phi)/r_o,$$

$$\gamma = \gamma_o/(1 + \varepsilon_1),$$

$$\gamma_o = (w'\cos\Phi - u'\sin\Phi)/\alpha_o - \sin(\Phi - \Phi_o). \tag{5.1.1}$$

Here, Φ_o and Φ are the angles of inclination of the material normal to the axis of rotation before and after deformation; κ_1 and κ_2 are the characteristics of the change of major curvatures; $\delta_k = 1 + \varepsilon_k$, $k = 1, 2, 3$. For the plate in the initial state, the angle of inclination of the normal to the axis of symmetry is $\Phi_o = 0$. The angle of the transverse shear γ is assumed to be small and is then considered to be zero.

In the dimensionless form, the differential system of equations for a plate loaded with uniform pressure has the form

$$d\tilde{T}^o/d\xi = \tilde{\alpha}_o \xi \delta_1 \delta_2 \tilde{p} \cos\Phi,$$

$$d\tilde{\Psi}^o/d\xi = \tilde{\alpha}_o \tilde{N}_2^o - \tilde{\alpha}_o \xi \delta_1 \delta_2 \tilde{p} \sin\Phi,$$

$$d\tilde{M}^o/d\xi = \tilde{\alpha}_o \tilde{M}_2^o \cos\Phi + \tilde{\alpha}_o \xi \delta_1 \tilde{Q}^o/\varepsilon_*,$$

$$d\tilde{w}/d\xi = \tilde{\alpha}_o \delta_1 \sin\Phi,$$

$$d\tilde{u}/d\xi = \tilde{\alpha}_o(\delta_1 \cos\Phi - 1),$$

$$d\Phi/d\xi = \tilde{\alpha}_o \tilde{K}_1, \tag{5.1.2}$$

where

$$\tilde{T}^o = \xi\tilde{V}^o, \quad \tilde{\Psi}^o = \xi\tilde{H}^o, \quad \tilde{M}^o = \xi\tilde{M}_1^o;$$

$$\tilde{V}^o = \tilde{N}_1^o \sin\Phi + \tilde{Q}^o \cos\Phi,$$

$$\tilde{H}^o = \tilde{N}_1^o \cos\Phi - \tilde{Q}^o \sin\Phi,$$

$$\tilde{\tilde{N}}_1^o = \tilde{V}^o \sin \Phi + \tilde{H}^o \cos \Phi,$$

$$\tilde{\tilde{Q}}^o = \tilde{V}^o \cos \Phi - \tilde{H}^o \sin \Phi; \tag{5.1.3}$$

$$\tilde{\tilde{N}}_1^o = \tilde{N}_1^o + \varepsilon_*(\tilde{K}_1 \tilde{M}_1^o + \tilde{K}_2 \tilde{M}_2^o)/\delta_1,$$

$$\tilde{\tilde{N}}_2^o = \tilde{N}_2^o + \varepsilon_*(\tilde{K}_1 \tilde{M}_1^o + \tilde{K}_2 \tilde{M}_2^o)/\delta_2. \tag{5.1.4}$$

Here, $\xi \in [0, 1]$ is independent of the radial Lagrangian coordinate; $\tilde{p}$ is the intensity of the hydrostatic pressure; $\varepsilon_* = h_*/R_*$ is a thin-walled parameter; $\tilde{V}^o$ and $\tilde{H}^o$ are internal efforts in the directions along the axis of symmetry and radius of the cylindrical coordinate system; and $\tilde{M}_1^o$ is the bending moment. The Lame coefficient $\tilde{\alpha}_o$ is assumed to be equal to one.

Let's recall the used material model and physical relations. Many metals and alloys behave almost as incompressible with large plastic deformations. So, the implemented condition is $\delta_1 \delta_2 \delta_3 = 1$.

The properties of the material are characterized by a loading diagram, which is approximated by a power function at the hardening site

$$\sigma = E\bar{e} \text{ when } \bar{e} \leq \bar{\varepsilon}_{0.2}; \quad \sigma = \bar{e}^\eta = E_s\bar{e}, \text{ when } e_{02} < \bar{e} < \bar{e}_l;$$

$$\bar{e} = (2/\sqrt{3})\sqrt{\bar{e}_1^2 + \bar{e}_1\bar{e}_2 + \bar{e}_2^2}, \tag{5.1.5}$$

where $\bar{e}$ is the intensity of logarithmic deformations of incompressible material; $\bar{e}_j = \ln(1+e_j)$; C, η are the material constants; $\sigma_{0.2}, \bar{\varepsilon}_{0.2}$ are stress and strain of conditional yield strength; E is Young's modulus; and $E_s = C\bar{e}^{\eta-1}$ is the secant modulus.

We use the physical ratio of the Davis–Nadai for an incompressible material, linking stress and the logarithmic strain in the principal axes as follows:

$$\sigma_1 = (4/3)\Lambda(\bar{e})(\bar{e}_1 + \nu\bar{e}_2), \quad \sigma_2 = (4/3)\Lambda(\bar{e})(\bar{e}_2 + \nu\bar{e}_1). \tag{5.1.6}$$

Here, $\Lambda = E$ at $\bar{e} \leq e_{0.2}$ (in the areas of elasticity) and $\Lambda(\bar{e}) = E_s(\bar{e})$ at $e_{0.2} < \bar{e} < \bar{e}_l$ (in plastic zones); $\tilde{\bar{e}} = \bar{e}/\bar{e}_l$, where $\sigma_l, \bar{e}_l$ is the maximum intensity of stresses and the maximum intensity of plastic deformations.

The chart of material properties is transferred into a single plane in the coordinates of the dimensionless stress $\tilde{\sigma} = \sigma/\sigma_l$ and the relative true strains $\tilde{\bar{e}} = \bar{e}/\bar{e}_l$ (see Chapter 2).

Further, to simplify the notation, the tildes ($\sim$) denoting a dimensionless value, are removed. When constructing two-dimensional equations, $\Lambda(\bar{e}) \approx \Lambda(\bar{\varepsilon})$. This is justified in the problem of strong drawing. The formulas for the connection of forces and moments with the components of deformations in a dimensionless form take the form

$$N_1^o = (k_\sigma/\varepsilon_*)B_1(\bar{\varepsilon}_1 + 0.5\bar{\varepsilon}_2),$$

$$N_2^o = (k_\sigma/\varepsilon_*)B_2(\bar{\varepsilon}_2 + 0.5\bar{\varepsilon}_1), \tag{5.1.7}$$

$$M_1^o = -k_\sigma D_1(\bar{K}_1 + 0.5\bar{K}_2),$$

$$M_2^o = -k_\sigma D_2(\bar{K}_2 + 0.5\bar{K}_1). \tag{5.1.8}$$

where

$$\bar{\varepsilon}_j = \ln(1 + \varepsilon_j) = \ln\delta_j, \quad \bar{K}_j = K_j/\delta_j, \quad j = 1,2, \quad k_\sigma = \sigma_l/E_*;$$

$$\bar{B}_1 = \bar{B}/\delta_1, \quad \bar{B}_2 = \bar{B}/\delta_2, \quad \bar{D}_1 = \delta_3^3\delta_2\bar{D}, \quad \bar{D}_2 = \delta_3^3\delta_1\bar{D},$$

$$\bar{B} = (4/3)\Lambda(\bar{\varepsilon})h_o, \quad \bar{D} = (1/9)\Lambda(\bar{\varepsilon})h_o^3. \tag{5.1.9}$$

Let's consider the pressure extraction of a flat circular plate of thickness h_p, pinched along the contour of the radius r_p, of an ellipsoidal dome (segment) of a given height w_o. When drawing in the vicinity of the ground loop, a sharp bend appears in the zone of which the flat pinched part of the plate passes into the dome. Experiments show that this area is very narrow (about 0.5 plate thickness). Therefore, its size can be neglected and taken for the dome radius of the support circuit equal to the radius of the original plate. This zone is conditionally cut off, and its force influence is replaced by reactions — vertical and horizontal linear forces as well as the moment. These quantities appear naturally when the equilibrium equations are integrated.

The meridian of the segment of the obtained shell corresponds to the arc of the ellipse with semi-axes a_e and b_e, the ratio $k_e = b_e/a_e$, and the square of eccentricity $e_x^2 = 1 - (k_e)^2$. This is true for $a_e > b_e$ (flattened ellipsoid). In the case of $a_e < b_e$ (elongated ellipsoid), $k_e = a_e/b_e$.

With a free plate extraction by hydrostatic pressure, the resulting dome passes the stages of both flattened and elongated ellipsoids of rotation. Previously, we have constructed a solution for only slightly flattened spheroids with the use of the smallness of the eccentricity that provides additional force on the apex. It is necessary to develop the method for the purposes of the extension of control over the form, identification of materials, and evaluation of limit states.

Let the ellipsoid be formed by rotating the ellipse around the z axis. We will consider two forms of the equation of an ellipse (the axial section of an ellipsoid). One of them is parametric,

$$r_1 = \phi(\tau) = a_e \sin \tau, \quad z = \psi(\tau) = z_e - b_e \cos \tau, \quad (5.1.10)$$

where z_c is a coordinate of the center of the ellipse. In this case, the polar coordinate r_1 is connected with the Lagrangian coordinate r, which is counted on the plate, through the radial component of the movement u: $r_1 = r + u$. Since the ellipsoids closed at the top are considered, then $\tau_0 = 0$. Parameter values τ on the path segment are denoted by τ_c. In the transition to dimensionless values, we believe that $h_* = h_p$, $R_* = r_p$, $E_* = \sigma_l$.

We define the τ_c parameters of the ellipse and the plate. On the contour of an elliptical segment, $r_1 = r_p$, $z = 0$. Then from (5.1.10), it follows that

$$\frac{r_p}{-z_e} = -\frac{a_e}{b_e} tg\tau_c = -\frac{1}{k_e} tg\tau_c, \quad \tau_c = arctg\left(\frac{k_e r_p}{z_e}\right). \quad (5.1.11)$$

The second form of geometry is classical in the axes r_1 and z of the cylindrical coordinate system.

$$(r_1/a_e)^2 + (z + z_c)^2/b_e^2 = 1. \quad (5.1.12)$$

Here, the radial coordinates r and r_1 vary from 0 to r_p. Dimensionless value $r_p = 1$.

By resolving (5.1.12) with respect to z, we have

$$z = f(r_1) = z_c - k_e\sqrt{a_e^2 - r_1^2}, \quad k_e = b_e/a_e. \quad (5.1.13)$$

In the parametric setting of the surface in the form of (5.1.10), its area is calculated by the integral

$$S = 2\pi \int_{\tau_0}^{\tau_c} \phi(\tau)\sqrt{(\phi_{,\tau})^2 + (\psi_{,\tau})^2}d\tau. \quad (5.1.14)$$

In the case of the form (5.1.10), after substituting the expressions for the derivatives of functions $\phi_{,\tau} = a_e \cos \tau$, $\psi_{,\tau} = b_e \sin \tau$, the integral (5.1.14) takes the form

$$S = 2\pi a_e \int_0^{\tau_c} F(\tau)d\tau, \tag{5.1.15}$$

where

$$F(\tau) = (\sin \tau)\sqrt{(a_e \cos \tau)^2 + (b_e \sin \tau)^2}. \tag{5.1.16}$$

To define a surface as (5.1.13), the area is given by the integral

$$S = 2\pi \int_0^{r_p} r_1 \sqrt{1 + (f(r_1),_{r_1})^2}dr_1. \tag{5.1.17}$$

Here, $f(r_1),_{r_1}$ is the derivative of the function. In explicit form,

$$f(r_1),_{r_1} = k_e r_1 / \sqrt{(a_e^2 - r_1^2)}. \tag{5.1.18}$$

We denote

$$\Pi(r_1) = r_1 \sqrt{1 + (f(r_1),_{r_1})^2}. \tag{5.1.19}$$

The volume of the shell with variable thickness $h(r_1)$, having an ellipsoid of rotation as the middle surface, can be obtained if the thickness is put under the signs of integrals (5.1.15), (5.1.17)

$$V_o = 2\pi a_e \int_0^{\tau_c} h(\tau)F(\tau)d\tau, \quad V_o = 2\pi \int_0^{r_p} h(r_1)\Pi(r_1)dr_1. \tag{5.1.20}$$

It is important to assume the form of the thickness distribution function. As in the previous chapter, we first use a quadratic dependence of the form

$$h(x) = h_p\{1 - \delta[1 - (x/x_c)^2]\}. \tag{5.1.21}$$

Subsequently, we will check other approximations. When choosing an independent argument, different options are possible here. The existing set of independent variables is the radii r and r_1, the parameter τ, the angle of inclination of the normal Φ, and the arc length of the curve l. For shells with small eccentricity, the best results are obtained from the dependence on Φ

$$h(\Phi) = h_p\{1 - \delta[1 - (\Phi/\Phi_c)^2]\}. \tag{5.1.22}$$

These parameters can be linked together. From (5.1.10), it follows that

$$\tau = arc\sin(r_1/a_e). \tag{5.1.23}$$

It can be shown that τ and Φ are not identical and are related by the following formulas:

$$\tau = arctg\left(\frac{a_e}{b_e}tg\Phi\right), \quad \Phi = arctg\left(\frac{b_e}{a_e}tg\tau\right). \tag{5.1.24}$$

These values can be identified only when $k_e = b_e/a_e$ is close to one.

Along with the angle of the normal Φ, it is also logical to use the length of the meridian arc as an independent coordinate. For the lengths of the arcs of the curve, the formulas are as follows:

$$L(\tau) = \int_0^\tau F_1(\tau)d\tau, \quad L(r_1) = \int_0^{r_1} \Pi_1(r_1)d\tau, \tag{5.1.25}$$

where

$$F(\tau) = \sqrt{(a_e \cos \tau)^2 + (b_e \sin \tau)^2}, \tag{5.1.26}$$

$$\Pi(r_1) = \sqrt{1 + (f(r_1)_{r_1})^2}. \tag{5.1.27}$$

Full length of meridian arc is $L_c = L(\tau_c) = L(r_p)$. Then, instead of the formula (5.1.21), we can take the following:

$$h(r_1) = h_p\{1 - \delta[1 - (L(r_1)/L_c)^2]\}. \tag{5.1.28}$$

Option (5.1.28) is more convenient for measurements in physical experiments.

For a one-parameter approximation of type (5.1.22), (5.1.28), the coefficient δ can be determined immediately, based on the incompressibility condition. The volume of the original plate having radius r_p and thickness h_p will be $V_p = \pi(r_p)^2 h_p$. Since the incompressible material is considered, then $V_o = V_p$. Applying the second of the formulas (5.1.20), we obtain

$$2\pi h_p \int_0^{r_p} \{1 - \delta[1 - (\Phi(r_1)/\Phi_c)^2]\}\Pi(r_1)dr_1 = \pi r_p^2 h_p. \tag{5.1.29}$$

The angle of inclination of the normal Φ is defined here as a function of r_1 via the formula $\tau = \arcsin(r_1/a_e)$ and (5.1.24).

We consider (5.1.29) as an equation to determine the coefficient δ. From this, it follows that

$$\delta = \frac{\left[\int_0^{r_p} \Pi(r_1)dr_1 - r_p^2/2\right]}{\int_0^{r_1} [1 - (\Phi(r_1)/\Phi_c)^2]\Pi(r_1)dr_1}. \qquad (5.1.30)$$

If the number of coefficients in the formula for thickness is more than one ($n_h > 1$), then the incompressibility condition will give the equation of connection between them. Then, $n_h - 1$ of these coefficients will become independent control parameters for the pressure functional constructed below.

Next, you can build a functional equation from which the radial displacement component is determined by an iterative process. To do this, the incompressibility condition is applicable to the areas of the dome and plate defined by the general Lagrangian coordinate r. There is an equation

$$\int_0^{r_1(r)} [h(r_1)\Pi(r_1)]dr_1 = r^2/2, \qquad (5.1.31)$$

where the variable is the upper limit $r_1(r) = r + u(r)$, and under the integral there is the integration variable. The independent variable in the equation (5.1.28) is the Lagrangian coordinate r. Here, we will not specify the type of thickness function.

We select a summand that can be easily integrated explicitly under the integral (5.1.28). To do this, add and subtract one

$$\int_0^{r_1(r)} [1 - 1 + h(r_1)\Pi(r_1)]dr_1 - r^2/2$$

$$= r + u(r) - \int_0^{r_1(r)} [1 - h(r_1)\Pi(r_1)]dr_1 - r^2/2. \qquad (5.1.32)$$

Hence,

$$u(r) = \int_0^{r_1(r)} [1 - h(r_1)\Pi(r_1)]dr_1 + r^2/2 - r. \qquad (5.1.33)$$

For the equation (5.1.33), it is possible to organize a simple iterative process on the basis of the principle of compressive maps, known from the functional analysis in [26]. Given some initial approximation $u_0(r)$ and calculating with it the right part in (5.1.30), we obtain the first approximation $u_1(r)$. Then the process is repeated, and for step $(k+1)$ we have

$$u_{k+1}(r) = \int_0^{r_{1k}(r)} [1 - h(r_1)\Pi(r_1)]dr_1 + r^2/2 - r, \qquad (5.1.34)$$

where $r_{1k}(r) = r + u_k(r)$. As an initial approximation, you can take the function

$$u_0(r) = k_u r(1 - r). \qquad (5.1.35)$$

The k_u coefficient is determined by the application of incompressibility conditions at the top of the dome and proximity to the uniformly deformed state of the middle surface: $\varepsilon_1(0) \approx \varepsilon_2(0)$. This gives

$$k_u = \frac{1}{\sqrt{1 - \delta}} - 1. \qquad (5.1.36)$$

The process (5.1.34) quickly converges. Five iterations are sufficient for accuracy of a few tenths of a percent. After exiting the process, the deformation components are determined $\varepsilon_2(\tau) = u(\tau)/\xi$, $\delta_2(\tau) = 1 + \varepsilon_2(\tau)$, $\delta_1(\tau) = [\delta_2(\tau)\delta_3(\tau)]^{-1}$, $\varepsilon_1(\tau) = \delta_1(\tau) - 1$, as well as internal efforts and moments (5.1.3), (5.1.4).

Then the system (5.1.2) is integrated. From the first, second, fourth and fifth equations, the following relations follow:

$$\bar{T}^\circ(\tau) = pJ_{11}(\tau) + P_0/(2\pi),$$

$$\bar{\Psi}^\circ(\tau) = J_{21}(\tau) - pJ_{22}(\tau) + r_p H_c; \qquad (5.1.37)$$

$$w(\xi) = -w_o + J_{41}(\xi), u(\xi) = J_{51}(\xi), \qquad (5.1.38)$$

where

$$J_{11}(\tau) = \int_0^\tau [\xi/\delta_3(\xi)] \cos \Phi(\xi) d\xi,$$

$$J_{21}(\tau) = \int_0^\tau \bar{N}_2^\circ(\xi) d\xi,$$

$$J_{22}(\tau) = \int_0^\tau [\xi/\delta_3(\xi)] \sin \Phi(\xi) d\xi,$$

$$J_{41}(\tau) = \int_0^\tau \delta_1(\xi) \sin \Phi(\xi) d\xi,$$

$$J_{51}(\tau) = \int_0^\tau \delta_1(\xi) \cos \Phi(\xi) d\xi. \qquad (5.1.39)$$

The displacements (5.1.35) vanish on the boundary contour. The values of P_0 and H_c are integration constants. In this case, P_0 makes sense concentrated at the top of the force affecting the shape of the shell and H_c is the radial force distributed along the contour.

From the third equation of the system (5.1.2) it is possible to determine the forming pressure of the dome of a given height w_0. This can be done in two ways: differential and integral. Since the geometry of the obtained shell is parametrically known, it is possible to determine the moments and their derivatives through its curvatures and obtained expressions of relative elongations. This is the first approach that gives the formula p the pressure:

$$p = \frac{[J_{21}(\tau) + r_p H_c] \sin \Phi(\tau) - [P_0/(2\pi)] \cos \Phi(\tau) + \varepsilon_* J(\tau)/\delta_1(\tau)}{J_{11}(\tau) \cos \Phi(\tau) + J_{22}(\tau) \sin \Phi(\tau)}$$

$$\approx const \qquad (5.1.40)$$

where

$$J(\tau) = M_1^o(\tau)\{1 - [\delta_1(\tau)/\delta_2(\tau)] \cos \Phi(\tau)\}$$
$$+ \tau[dM_1^o(\tau)/d\tau]. \qquad (5.1.41)$$

If we integrate the third equation of the system, we will obtain the formula

$$p = \frac{\tau M_1^o(\tau) + r_p M_c - J_{31}(\tau) - J_{33}(\tau) + J_{34}(\tau) + J_{36}(\tau)}{J_{32}(\tau) + J_{35}(\tau)}$$

$$\approx const. \qquad (5.1.42)$$

Here,

$$J_{31}(\tau) = \int_0^\tau M_2^o \cos \Phi(\xi) d\xi,$$

$$J_{33}(\tau) = \frac{P_0}{2\pi\varepsilon_*} \int_0^\tau \delta_1(\xi) \cos \Phi(\xi) d\xi,$$

$$J_{34}(\tau) = \frac{1}{\varepsilon_*} \int_0^\tau \delta_1(\xi) J_{21}(\xi) \sin \Phi(\xi) d\xi,$$

$$J_{36}(\tau) = \frac{H_c}{\varepsilon_*} \int_0^\tau \delta_1(\xi) \sin \Phi(\xi) d\xi,$$

$$J_{32}(\tau) = \frac{1}{\varepsilon_*} \int_0^\tau \delta_1(\xi) J_{11}(\xi) \cos \Phi(\xi) d\tau,$$

$$J_{35}(\tau) = \frac{1}{\varepsilon_*} \int_0^\tau \delta_1(\xi) J_{22}(\xi) \sin \Phi(\xi) d\xi. \qquad (5.1.43)$$

The output of the expression (5.1.40) by a constant is controlled by the parameters P_0, H_c, e_x. In (5.1.42), these values are added to the constant M_c, which appears when integrating the third equation. A larger number of control parameters makes it more advantageous to apply the formula (5.1.42), especially when forming high lift shells.

In constructing the solution for thickness, the formula (5.1.28) was mainly used. A generalized solution was tested by comparison with the more analytical version, built for a slightly flattened ellipsoid [16, 19].

Along with the basic formula for approximating the thickness function (option 1), other options with additional terms were also considered to improve the accuracy of the functional (5.1.42) reaching the stationary value. Options 2–8 for thickness are presented below.

$$h_2(r_1) = h_o\{1 - \delta[1 - F_{lc}(r_1)^2] - g_h e_x^2[1 - F_{lc}(r_1)]^2\};$$

$$h_4(r_1) = h_o\{1 - \delta[1 - F_{lc}(r_1)^2] - g_k e_x^2[1 - F_{lc}(r_1)]\};$$

$$h_3(r_1) = h_o\{1 - \delta[1 - F_{lc}(r_1)] - g_g e_x^2[1 - F_{lc}(r_1)^4]\};$$

$$h_5(r_1) = h_o\{1 - \delta[1 - F_{lc}(r_1)^2] - g_3 e_x^2[1 - F_{lc}(r_1)^3]$$
$$- g_4 e_x^2[1 - F_{lc}(r_1)^4]\};$$

$$h_6(r_1) = h_o\{1 - \delta[1 - F_{lc}(r_1)^2] - g_1 e_x^2[1 - F_{lc}(r_1)]$$
$$- g_3 e_x^2[1 - F_{lc}(r_1)^3] - g_4 e_x^2[1 - F_{lc}(r_1)^4]\};$$

$$h_7(r_1) = h_o\{1 - \delta[1 - F_{lc}(r_1)^2] \cdot [1 - g_e e_x^2 F_{lc}(r_1)]\};$$

$$h_8(r_1) = h_o\{1 - \delta[1 - F_{lc}(r_1)^2] \cdot [1 - g_e e_x^2 F_{lc}(r_1)]$$
$$- g_e e_x^2[1 - F_{lc}(r_1)]^2\}, \tag{5.1.44}$$

where $F_{lc}(r_1) = L(r_1)/L_c$ from formula (5.1.28).

These options allow to improve the accuracy of the functional output to a stationary value from 1–1.5% to 0.5–0.25%. As shown by numerical experiments, the fifth option is the most preferable.

The model equations and the calculation process are performed in dimensionless form. In particular, in the normalization of displacement and pressure, the following formulas are used:

$$\{u, w\}_U = \{u, w\}_D/R_*, \quad \{p\}_U = \{p\}_D/(E_* \varepsilon_*), \quad \varepsilon_* = h_*/R_*. \tag{5.1.45}$$

Here, u, w are horizontal and vertical displacements; h_*, R_* are characteristic small and large sizes; ε_* is a thin-walled parameter; and indices "U" and "D" are marked, respectively, dimensionless and dimensional quantities. As characteristic values were taken $h_* = h_p$, $R_* = r_p$, $E_* = E$ (Young's modulus) or σ_l (tensile strength). The calculation results can be output both in dimensionless and dimensional forms.

Let us present some calculation results in dimensional form, to return to which formulas (5.1.45) are used. We consider a plate with thickness $h_p = 0.38\,mm$ and radius of the reference circuit $r_p = 100\,mm$, Young's modulus $E = 0.21 \cdot 10^6\,MPa$ (megapascals), yield strength $\sigma_{02} = 360\,MPa$ tensile strength (sigma temporary) $\sigma_l = 720\,MPa$, ultimate strain intensity $\bar{\varepsilon}_l = 0.615$, strain strength of flow $\varepsilon_{02} = \sigma_{02}/E = 0.001714$, and $\varepsilon_* = 0.0038$. The diagram of the plastic properties of the material corresponds to stainless steel 12X18H10T. Such plates are used for the manufacture of the destructible elements of the devices that protect technological equipment and tanks from destruction by excess pressure [56].

When the shell is drawn freely by pressure without applying force at the top, it passes through the stages of segments of a slightly flattened ellipsoid, a spherical dome, and an elongated ellipsoid.

Figure 5.1.1 shows the nonlinear dependence of the molding pressure on the height of the dome $p_f - w_o$, the graph of which has the form observed in physical experiments. Pressure is given in MPa, dome height in mm.

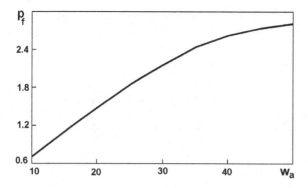

Fig. 5.1.1. The dependence of the forming pressure on the dome height.

Table 5.1.1. The relationship between dome height and eccentricity.

w_0 (mm)	0.10	0.15	0.20	0.25	0.30	0.35	0.40	0.45	0.50
e_x		0.15	0.007	0.00	0.16	0.20	0.21	0.22	0.23	0.25

In the Table 5.1.1 the compliance values of the height of the dome w_a and the eccentricity e_x of the ellipsoid shell meridian are presented.

A more general form of the constructed algorithm makes it possible to consider options for ellipsoids with fairly large eccentricities. The solution, where the power series expansion by eccentricity is used, follows from the generalized one as a special case.

The height is presented in a dimensional form (in *mm*).

Although for more efficient computability this technique can be used for small eccentricities, a small amount of eccentricity is not assumed in the generalized approach presented here. This makes it possible to expand the possibilities of applying this semi-inverse method in the case of deformations of plastic materials close to the limiting ones.

5.2. Inflating of a Cylindrical Shell

We consider a semi-inverse method for solving problems of deformation of shells of rotation. Here, we investigate the possibility of using analytical approximations in the problem of inflating a cylindrical shell clamped at the ends [52].

The equations of the problem use kinematic relations of the E. Reissner type, generalized to large elongations and taking into account the compression of the normal

$$e_1 = (\varepsilon_1 + \zeta\kappa_1), \quad e_2 = (\varepsilon_2 + \zeta\kappa_2),$$

$$\varepsilon_1 = (w'\sin\Phi + u'\cos\Phi)/\alpha_o + \cos(\Phi - \Phi_o) - 1, \quad \varepsilon_2 = u/r_o$$

$$\kappa_1 = \Phi'_o/\alpha_o - (1+\varepsilon_3)K_1, \quad \kappa_2 = (\sin\Phi_o)/r_o - (1+\varepsilon_3)K_2$$

$$K_1 = \Phi'/\alpha_o, \quad K_2 = (\sin\Phi)/r_o; \tag{5.2.1}$$

$$\gamma_1 = \gamma_o/(1+\varepsilon_1),$$

$$\gamma_o = (w'\cos\Phi - u'\sin\Phi)/\alpha_o - \sin(\Phi - \Phi_o); \tag{5.2.2}$$

$$u' = \alpha_o[\varepsilon_1\cos\Phi - \gamma_o\sin\Phi + (\cos\Phi - \cos\Phi_o)],$$

$$w' = \alpha_o[\varepsilon_1\sin\Phi + \gamma_o\cos\Phi + (\sin\Phi - \sin\Phi_o)]. \tag{5.2.3}$$

Here, Φ_o and Φ are the angles of inclination of the material normal to the axis of rotation before and after deformation; and κ_1 and κ_2 are characteristics of change of the main curvatures. The positive direction of the normal is inside the shell. All variables are functions of the longitudinal Lagrangian curvilinear coordinate. This is the axial coordinate z for a cylindrical shell. Tilt angle for it is $\Phi_o = \pi/2$, $\alpha_o(z) = 1$, $r_o = r_c$, where r_c is the radius of the middle surface of the cylindrical shell.

In the transition from a three-dimensional body to a shell model in the virtual work of internal forces, the integral characteristics of the stress state are introduced ($\zeta \in [-h_o/2, h_o/2]$) as follows:

$$N_1^o = (1+\varepsilon_3)\int_{(h_o)}\sigma_1(1+e_2)d\zeta,$$

$$N_2^o = (1+\varepsilon_3)\int_{(h_o)}\sigma_2(1+e_1)d\zeta,$$

$$M_1^o = (1+\varepsilon_3)^2\int_{(h_o)}\sigma_1(1+e_2)\zeta d\zeta,$$

$$M_2^o = (1+\varepsilon_3)^2\int_{(h_o)}\sigma_2(1+e_1)\zeta d\zeta,$$

$$Q^o = (1+\varepsilon_3)\int_{(h_o)}[\sigma_{13} + \gamma\sigma_1(1+e_2)]d\zeta. \tag{5.2.4}$$

The equilibrium equations follow from the principle of possible displacements as

$$(r_o \bar{V}^o)' + \alpha_o r_o p_w^o = 0,$$
$$(r_o \bar{H}^o)' - \alpha_o \bar{N}_2^o + \alpha_o r_o p_u^o = 0,$$
$$(r_o M_1^o)' - \alpha_o M_2^o \cos \Phi + \alpha_o r_o [\gamma_o \bar{N}_1^o - (1 + \varepsilon_1) Q^o] = 0, \quad (5.2.5)$$

where

$$\bar{V}^o = \bar{N}_1^o \sin \Phi + Q^o \cos \Phi,$$
$$\bar{H}^o = \bar{N}_1^o \cos \Phi - Q^o \sin \Phi,$$
$$p_w^o = \delta_1 \delta_2 p_w, \quad p_u^o = \delta_1 \delta_2 p_u,$$
$$\delta_1 = 1 + \varepsilon_1, \quad \delta_2 = 1 + \varepsilon_2.$$

Values $\bar{V}^o$ and $\bar{H}^o$ have the meaning of internal forces oriented, respectively, along the axis of symmetry and the radius of the cylindrical coordinate system; p_u, p_w are components of the external surface of the servo load. Here, the transverse shear angle γ_o is considered small. Subsequently when forming physical relationships, it is assumed to be zero. We also use the assumption of incompressibility of the material, which works well for metals and alloys. For an incompressible material,

$$\bar{N}_1^o = N_1^o + (K_1 M_1^o + K_2 M_2^o)/\delta_1,$$
$$\bar{N}_2^o = N_2^o + (K_1 M_1^o + K_2 M_2^o)/\delta_2.$$

Material properties are characterized by a loading diagram approximated by a power function

$$\sigma = C\bar{e}^\eta = \Lambda(\bar{e})\bar{e}, \quad \bar{e} = (2/\sqrt{3})\sqrt{\bar{e}_1^2 + \bar{e}_1 \bar{e}_2 + \bar{e}_2^2},$$

where $\bar{e}$ is the intensity of logarithmic deformations of incompressible material; $\bar{e}_k = \ln(1 + e_k)$; C, η are the material constants; and $\Lambda(\bar{e}) = C\bar{e}^{\eta-1}$ is the secant modulus.

For an incompressible material, we use the physical ratio of the Davis–Nadai linking stress and logarithmic strain

$$\sigma_1 = (4/3)\Lambda(\bar{e})(\bar{e}_1 + \nu \bar{e}_2), \quad \sigma_2 = (4/3)\Lambda(\bar{e})(\bar{e}_2 + \nu \bar{e}_1). \quad (5.2.6)$$

After substitution (5.2.6) in (5.2.4), the defining relations for the generalized internal forces are constructed as follows. The function $\Lambda(\bar{e})$ is replaced by the sum of several terms of the Taylor power series by ζ in the neighborhood $\zeta = 0$. A margin three terms of approximation in the transverse coordinate is sufficient. After integration, the degrees h_o are retained not above the third. As a result, nonlinear relations are obtained that connect forces, moments, strains $\varepsilon_k = \bar{e}_k|_{z=0}$, $\bar{\varepsilon}_k = \ln(1 + \varepsilon_k)$, $k = 1, 2, 3$, and the parameters change of the curvatures κ_1, κ_2.

In the simplest version of the defining relations, we can put the secant modulus i.e. replace it with the corresponding dependence on the middle surface. This is justified for the problems of strong drawing and is equivalent to neglecting the variability of the material properties in thickness. In addition, we assume that in the region under consideration in the resulting shell does not have zones of strong bending: $\max\{\zeta\bar{\kappa}_1, \zeta\bar{\kappa}\} \ll 1$. Strong regional strip bending is eliminated. As a result, we have the following nonlinear physical relations:

$$N_1^{\mathrm{o}} = \bar{B}_1(\bar{\varepsilon}_1 + 0.5\bar{\varepsilon}_2), \quad N_2^{\mathrm{o}} = \bar{B}_2(\bar{\varepsilon}_2 + 0.5\bar{\varepsilon}_1),$$
$$M_1^{\mathrm{o}} = \bar{D}_1(\bar{\kappa}_1 + 0.5\bar{\kappa}_2), \quad M_2^{\mathrm{o}} = \bar{D}_2(\bar{\kappa}_2 + 0.5\bar{\kappa}_1), \qquad (5.2.7)$$

where

$$\bar{\kappa}_1 = \kappa_1/\delta_1, \quad \bar{\kappa}_2 = \kappa_2/\delta_2, \quad \bar{B}_1 = \bar{B}/\delta_1,$$
$$\bar{B}_2 = \bar{B}/\delta_2, \quad \bar{D}_1 = \delta_3^2\delta_2\bar{D}, \quad \bar{D}_2 = \delta_3^2\delta_1\bar{D}. \qquad (5.2.8)$$

In the initial state, the cylindrical shell has a constant thickness h_o, diameter d_c, radius $r_o = r_c = d_c/2$, and length $L_c = 0.2$. We denote the half length by $L_{05} = L_c/2$. Considering, for convenience, the z axis of the circular symmetry of the horizontal cylinder, we place the origin in its middle. Then, the coordinates of the ends correspond to the values $z = -L_{05}$ and $z = L_{05}$. In view of the symmetry relative to the middle, half of the structure can be considered. The second coordinate is measured by the radius r. The ends of the shell are assumed to be fixedly clamped. Inside the shell, a hydraulic pressure is applied, as a result of which the shell is plastically deformed and takes a barrel shape, Fig. 5.2.1. The ordinates of the meridian of the shell in the deformed state are determined by the function $r_1(\zeta)$.

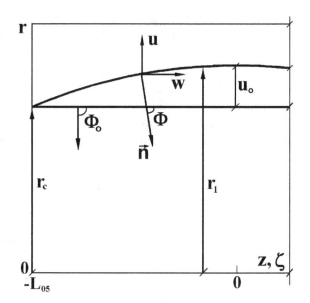

Fig. 5.2.1.   Coordinate system.

Let us explain the difference between the coordinates z and ζ, counted on the axis of rotation of the shells (Fig. 5.2.1). The z coordinate is Lagrangian, which determines the position of the material point on the meridian of the shell. It is also an independent variable in the equations of the mathematical model. The coordinate ζ is Eulerian and defines the geometric points through which the meridian of the deformed shell passes when it intersects with the radial beam, if it is drawn from the point with the coordinate z. The coordinates z and ζ differ by the axial displacement w of the material point: $\zeta = z + w$. After defining $w(z)$, you can return to the Lagrangian coordinates when you process the equations of the mathematical model by specifying ζ as a function of z: $\zeta(z) = z + w(z)$.

With large deformations, the initially constant thickness of the shell becomes variable along the axial coordinate due to transverse compression (compression of the material normal). Consideration of this factor is significant, affecting the elongation in tangential directions. Since the shell material behaves as incompressible and retains the original density, it is possible to distribute the initial mass of the material in the same volume when moving to a shell with an increased area of the middle surface only by reducing the thickness.

The thickness has a maximum at the clamping points and varies from the initial value to the minimum at the point of greatest displacement u_0. When drawing a dome of average height, the thickness distribution is subject to a quadratic dependence on the angle of inclination of the normal (or on the arc length, Sections 4.3, 4.4, 5.1). This is confirmed both theoretically and experimentally.

Based on this result and the similarity of the Meridian shapes, for the cylinder, we also assume a quadratic dependence, but in the coordinate ζ

$$h(\zeta) = h_0[1 - \delta(1 - \zeta^2)]. \tag{5.2.9}$$

The coefficient δ can be determined from the condition of equality of volumes V_0 and V_1 of the shell in the initial and deformed states

$$V_0 = 2\pi r_c h_0 L_{05},$$

$$V_1 = 2\pi \int_{-L_{05}}^{0} r_1(\zeta)\alpha_0(\zeta)\, h(\zeta)\, d\zeta,$$

$$2\pi \int_{-L_{05}}^{0} r_1(\zeta)\alpha_0(\zeta) h(\zeta) d\zeta = 2\pi r_c h_0 L_{05}. \tag{5.2.10}$$

Substituting (5.2.9) in (5.2.10), we have

$$\int_{-L_{05}}^{0} r_1(\zeta)\alpha_0(\zeta)[1 - \delta(1 - \zeta^2)]d\zeta = r_c L_{05}. \tag{5.2.11}$$

From (5.2.11), after integration the formula follows that

$$\delta = (J_1 - r_c L_{05})/(J_1 - J_2), \tag{5.2.12}$$

where

$$J_1 = \int_{-L_{05}}^{0} r_1(\zeta)\alpha_0(\zeta)d\zeta, \quad J_2 = \int_{-L_{05}}^{0} r_1(\zeta)\alpha_0(\zeta)\zeta^2 d\zeta. \tag{5.2.13}$$

For the possibility of adjustment, we add a fourth degree, setting the thickness function also in the form

$$h(\zeta) = h_0[1 - \delta(1 - \zeta^2) - \delta_1(1 - \zeta^4)]. \tag{5.2.14}$$

The coefficients δ and δ_1 can be linked by the ratio (5.2.15), applying the condition of equality of volumes V_0 and V_1, which gives

$$\delta_1 = [J_1 + \delta(J_4 - J_1) - r_c L_{05}]/(J_1 - J_4),\qquad(5.2.15)$$

where

$$J_4 = \int_{-L_{05}}^{0} r_1(\zeta)\alpha_o(\zeta)\zeta^4 d\zeta,$$

and J_1 is the same as in (5.2.12), (5.2.13). Here, δ is set as the main parameter through which δ_1 is defined.

We note right away that numerical experiments have not shown the advantages of a more complex formula (5.2.14) compared with (5.2.9). Thus, a quadratic approximation of thickness is also acceptable here.

To determine the axial movement of points $w(z)$, we again apply the incompressibility condition, but already to the variable volume of the original and deformed shells, cut off by Parallels with the current coordinate z: $V_z(\zeta(z)) - V_{0z}(z) = 0$, or

$$\int_{-L_{05}}^{\zeta(z)} [r_1(\xi)\alpha_o(\xi)h(\xi)]d\xi - h_o(z + L_{05}) = 0. \qquad(5.2.16)$$

We add and subtract constant h_o in expanded expression (5.2.16), partially integrate and substitute $\zeta(z) = z + w(z)$

$$h_o[z + w(z) + L_{05}]$$

$$= h_o(z + L_{05}) + \int_{-L_{05}}^{z+w(z)} [h_o - r_1(\zeta)\alpha_o(\zeta)h(\zeta)]d\zeta. \qquad(5.2.17)$$

From (5.2.17), it follows that

$$w(z) = \frac{1}{h_o} \int_{-L_{05}}^{z+w(z)} [h_o - r_1(\zeta)\alpha_o(\zeta)h(\zeta)]d\zeta. \qquad(5.2.18)$$

Using the relation (5.2.18), we construct a simple iterative process based on the principle of compressed maps [26]

$$w_k(z) = \frac{1}{h_o} \int_{-L_{05}}^{z+w_{k-1}(z)} [h_o - r_1(\zeta)\alpha_o(\zeta)h(\zeta)]d\zeta. \qquad(5.2.19)$$

The function $w(z)$ vanishes at points $z = -L_{05}$ and $z = 0$, less than zero within this interval, and has one extreme (minimum). These conditions are satisfied by the function

$$w_0(z) = W_0 z(1 + z/L_{05}), \qquad (5.2.20)$$

which we take as zero approximation. The iterative process (5.2.19) converges quickly. One or two iterations are enough, and the second iteration is necessary to confirm the accuracy of the first. Moreover, if after the first iteration we choose W_0 in (5.2.20) so that the minima coincide, then this approximation is already a very good approximation, and it can be used in further calculations.

After determining $w(z)$, all characteristics of the shell in the deformed state can be written in the Lagrangian coordinates, which are used in the equilibrium equations. The displacement vector and the deformation field become known if the incompressibility condition is used locally at the points of the middle surface

$$[1 + \varepsilon_1(z)] \cdot [1 + \varepsilon_2(z)] \cdot [1 + \varepsilon_3(z)] = 1. \qquad (5.2.21)$$

We have the radial displacement component $u(z) = r_1(\zeta(z)) - r_c$; the circumferential elongation $\varepsilon_2(z) = u(z)/r_c$; and the normal compression (relative thickness change)

$$\varepsilon_3(z) = [h(\zeta(z)) - h_o]/h_o] < 0.$$

The elongation at the meridian follows from (5.2.21) due to local incompressibility: $\varepsilon_1(z) = [\delta_2(z)\delta_3(z)]^{-1} - 1$, where $\delta_j(z) = (1 + \varepsilon_j(z))$, $j = 1, 2, 3$. The thickness of the deformed shell is $h(z) = h_o\delta_3(z)$. The intensity of the logarithmic deformation of the middle surface is

$$\bar{\varepsilon}(z) = (2/\sqrt{3})\sqrt{\bar{\varepsilon}_1(z)^2 + \bar{\varepsilon}_1(z) \cdot \bar{\varepsilon}_2(z) + \bar{\varepsilon}_2(z)^2},$$

where

$$\bar{\varepsilon}_j(z) = \ln(1 + \varepsilon_j(z)) = \ln(\delta_j(z)).$$

We introduce notations for the functions of the angles of inclination of the tangent and normal

$$f_1(\zeta) = -arctg(r_1(\zeta)), \quad f(\zeta) = f_1(\zeta) + \pi/2,$$

$$\Phi(z) = f(\zeta(z)), \quad \Phi_1(z) = \pi/2 - \Phi(z).$$

By integrating the first two equilibrium equations from (5.2.5) and substituting the necessary relations into the third, a formula for uniform pressure p can be obtained

$$p = G_1(z)/G_2(z) \approx const, \qquad (5.2.22)$$

where

$$G_1(z) = [-(r_o M_1^o)' + M_2^o(z) \cos \Phi(z)] \delta_1^{-1}(z)$$
$$+ I_{21}(z) \sin \Phi(z) + C_H \sin \Phi(z) - C_V \cos \Phi(z),$$

$$G_2(z) = I_{11}(z) \cos \Phi(z) + I_{22}(z) \sin \Phi(z),$$

$$I_{11}(z) = \int_{-L_{05}}^{z} \delta_1(\xi) \delta_2(\xi) \sin \Phi_1(\xi) d\xi, \quad I_{21}(z) = \int_{-L_{05}}^{z} \bar{N}_2^o(\xi) d\xi,$$

$$I_{22}(z) = \int_{-L_{05}}^{z} \delta_1(\xi) \delta_2(\xi) \cos \Phi_1(\xi) d\xi. \qquad (5.2.23)$$

The value of p must be constant. The task of the semi-inverse method is to select the parameters of the deformed shell such as to satisfy this condition. The formula (5.2.22) also includes the integration constants C_V and C_H, which make sense of axial and radial reactions corresponding to the clamping of the ends. They are also the right-hand side view control parameters in the formula (5.2.22).

We consider a shell, which is cylindrical, of constant thickness in the initial state, and has dimensions (in m): the diameter of $d_c = 0.2$, the radius $r_o = r_c = d_c/2 = 0.1$, the thickness $h_o = 3.8 \cdot 10^{-3}$, length $L_c = 0.2$, half of the length $L_{05} = L_c/2 = 0.1$. The dimensions correspond to the side projection, which fits into a square with a side of $0.2\, m$.

As the material of the shell take stainless steel 12X18H10T, since it has good properties of plasticity. Characteristic parameters of the loading diagram have values (stresses in MPa): Young's modulus $E = 2.1 \cdot 10^5$, elastic limit $\sigma_{02} = 360$, the limit of elastic deformation $\varepsilon_{02} = \sigma_{02}/E = 1.7143 \cdot 10^{-3}$, tear stress intensity $\sigma_l = 720\, MPa$, limit intensity of deformations $\bar{\varepsilon}_l = 0.615$. According to these data, the parameters of the power approximation of the diagram of nonlinear properties of the material are determined

$$\sigma(\bar{\varepsilon}) = C\bar{\varepsilon}^\eta, \quad C = \sigma_l/\bar{\varepsilon}^\eta = 762.45,$$

$$\eta = [\ln(\sigma_l) - \ln(\sigma_{02})]/[\ln(\bar{\varepsilon}_l) - \ln(\varepsilon_{02})] = 0.1178.$$

The secant modulus is $\Lambda(\bar{\varepsilon}) = C\bar{\varepsilon}^{\eta-1}$.

In the transition to dimensionless, as the normalizing voltage we take σ_l. The displacements are normalized by the radius of the original shell.

For example, we will set the arrow of the greatest displacement $u_b = 0.035\,m$, which is 35% of the original radius. Then the greatest radius of the shell in the deformed state will be $r_b = 0.135\,m$. Respectively, in the dimensionless form $\tilde{r}_c = 1.0$, $\tilde{u}_b = 0.35$, $\tilde{r}_b = 1.35$, $\tilde{L}_{05} = 1.0$.

After the transition to dimensionless quantities, smooth curves pass through the boundary points $(\mp\tilde{L}_{05}, \tilde{r}c)$ and the maximum displacement point $(0, \tilde{r}_b)$. Functions that define the ordinates of the meridian depending on the axial variable $\tilde{z}$ should be well differentiated to ensure a smooth curvature. The following functions were taken as such:

(1) Power function

$$\tilde{f}_p(\tilde{z}) = \tilde{r}_b - \tilde{u}_b|\tilde{z}|^s,$$

where s varies and can be non-integer; at $s = 2$, the function turns into a parabola.

(2) The trigonometric cosine in degrees

$$\tilde{f}_c(\tilde{z}) = \tilde{A}_c[\cos(N_c\tilde{z})]^s + \tilde{B}_c,$$

where

$$\tilde{A}_c = -\tilde{u}_b/\{[\cos(N_c\tilde{L}/2)]^s - 1\}, \quad \tilde{B}_c = \tilde{r}_b - \tilde{A}_c,$$

n_c and s vary.

(3) Hyperbolic cosine in degree

$$\tilde{f}_h(\tilde{z}) = \tilde{A}_h[\cos(N_h\tilde{z})]^s + \tilde{B}_h,$$

where

$$\tilde{A}_h = -\tilde{u}_b/\{[\text{ch}(N_h\tilde{L}/2)]^s - 1\}, \tilde{B}_h = \tilde{r}_b - \tilde{A}_c,$$

n_h and s vary.

(4) Circular arc

$$\tilde{f}_o(\tilde{z}) = \sqrt{R_o - \tilde{z}^2} - \tilde{B}_o,$$

where $\tilde{R}_o = \tilde{r}_b + \tilde{B}_o$ is the radius of the arc,

$$\tilde{B}_o = [(\tilde{L}/2)^2 - 2\tilde{r}_c \tilde{u}_b - \tilde{u}_b^2]/(2\tilde{u}_b).$$

The function is uniquely determined because the arc passes through three points.

(5) The arc of the ellipse

$$\tilde{f}_e(\tilde{z}) = k_e \sqrt{a_e^2 - \tilde{z}^2} - \tilde{B}_e,$$

where a_e, b_e are semi-axes of ellipse, $k_e = b_e/a_e$,

$$\tilde{B}_e = [(2a_e/\tilde{L})^2 - 1]\tilde{u}_b - \tilde{r}_c + (2a_e/\tilde{L})\tilde{u}_b \sqrt{1 - (2a_e/\tilde{L})^2}.$$

Form management options are a_e, k_e.

(6) Combinations of functions with weights

option 1 – the mix of functions (5.2.2) and (5.2.5)

$$\tilde{f}_1(\tilde{z}) = m_1 \tilde{f}_c(\tilde{z}) + (1 - m_1)\tilde{f}_o(\tilde{z});$$

option 2 — the mix of functions (5.2.2) and (5.2.1)

$$\tilde{f}_2(\tilde{z}) = m_2 \tilde{f}_c(\tilde{z}) + (1 - m_2)\tilde{f}_p(\tilde{z});$$

option 3 — the mix of functions (5.2.2), (5.2.1), and (5.2.5)

$$\tilde{f}_2(\tilde{z}) = m_1 \tilde{f}_c(\tilde{z}) + m_2 \tilde{f}_p(\tilde{z}) + m_3 \tilde{f}_o(\tilde{z}),$$

where $m_1 + m_2 + m_3 = 1$.

Since each of the functions (5.2.1)–(5.2.5) passes through the given points, their weight combinations also satisfy these conditions. Functions with logarithms and some other combinations were also considered.

The algorithm for selecting suitable approximations can be as follows. For each of the test functions (5.2.1)–(5.2.5) first, their parameters are selected, which give the best possible approximation of the function (5.2.22) to the constant. Then two or three most suitable functions are selected, which are combined in various forms.

Consider, for example, stretching of the shell to the value of displacement in the middle $\tilde{u}_o = 0.35$; respectively, $\tilde{r}_1 = 1.35$. The selection of test function parameters gives the following values:

$$(1) \ s = 1.5\text{--}1.8; \quad (2) \ s = 1.5, \quad N_c = 1.5, \quad \tilde{A}_c = 0.3567,$$

$$\tilde{B}_c = 0.9933; \quad (3) \ s = 1.2, \quad N_h = 0.5, \quad \tilde{B}_h = 2.3627,$$

$$\tilde{A}_h = -1.0127; \quad (4) \ \tilde{B}_o = 0.2536, \quad \tilde{R}_o = 1.6036.$$

A comparison of variants shows that the most acceptable is variant 3 of (5.2.6) — a mix of cosine, power function, and arc of a circle with weight coefficients $m_1 = 0.4$, $m_2 = 0.25$, $m_3 = 0.35$. In the central part of the region ($\tilde{z} \in [-0.95, 0.1]$), the error of pressure output to the constant is 4–6% deviation from the mean value $p_{mid} \approx 1.7\,MPa$. At the edges, the error is greater ($\sim14\%$). However, control by solving the Cauchy problem for a system of sixth-order equations constructed on the basis of relations (5.2.1)–(5.2.3), (5.2.5)–(5.2.7) with the initial conditions on the left edge of the semi-inverse method gives an almost exact match to the matched geometry.

Chapter 6

About Containers for Liquid Cargo Transportation

The globalization of the economy and the expansion of markets are inevitably associated with an increase in cargo turnover and there is also the associated geography and distances of transportation. Among the transported goods, a considerable tonnage is occupied by environmentally hazardous and harmful-to-human-health liquids and substances. For their transportation, containers such as barrels are widely used, Fig. 6.1.1. The significance of the danger is related to the properties of the transported goods. Liquids and substances can have a low boiling point, a tendency to heat up when combined with air, cause fire or explosion, be poisonous when inhaled and ingested by living tissues, etc.

6.1. Problems of Containers Strength

Features of the tank body are as follows [46–49]. During the transportation of liquid materials, the tank experiences dynamic loads due to fluctuations in the vehicle during transportation as well as in acceleration and braking modes. The resulting stresses are summed up with stresses caused by heating or cooling, excessive or negative internal pressure.

Multiple variable loads lead to fatigue cracks and a breach of tightness, Figs. 6.1.2 and 6.1.3. Due to the specific nature of the

Fig. 6.1.1. Different types of barrels.

substances transported, this causes serious environmental and economic damage.

In solving the complex emerging problems, an important task is to establish a chain of cause-and-effect relationships that leads

Fig. 6.1.2. Versions of the cracks.

Fig. 6.1.3. Deformations and cracks formed when capacity drops.

to damage to the shells of dangerous substances. If the tightness of the tanks is broken, then there are problems of improving their designs.

The crack on the bottom of the container arises as a point defect, Fig. 6.4.2. Then, there is its growth and development in various forms and orientations. Figure 6.4.3 shows a cruciform crack. Other possible options are given in Fig. 6.4.1.

Attention to the topic arose in the framework of creative cooperation with the University of Dortmund, where the test cycle of works on tanks was carried out. The natural development of this topic was the mathematical modeling of the complex of problems. The development of methods and approaches in this area has been implemented by us in Southern Federal University.

6.2. Parameterization of the Bottom of a Composite Geometry

A typical container considered in the research is shown in Fig. 6.2.1. The design of the container (barrel, tank) is related to the manufacturing technology and the need to ensure tightness. At the junction of the bottom with the cylindrical shell, there is a rib formed by wrapping the edges using a sealant. The transition to the edge from the plate goes through a narrow zone of great curvature of the Meridian, called the shoulder.

The lower bottom is the weakest element of the structure, which is due to its configuration and the highest load. The weight action of the internal liquid in a vertically standing cylindrical container, closed at the ends by plates or flat shells, creates the greatest pressure on the bottom. Studies have shown that increased stress can lead to fatigue cracks and destruction in this element of the composite shell of the container under dynamic influences that overlap with the static stress state. In this regard, it is advisable to analyze, first of all, models for the bottom loaded with the weight of the liquid in

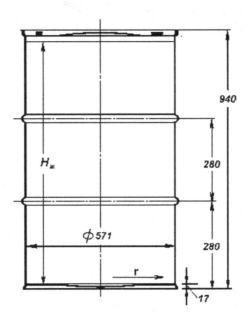

Fig. 6.2.1.   Sketch of a typical container.

the filled container. At the same time, it is necessary to investigate both a static problem and a dynamic one, taking into account static stresses and the reaction of the internal environment. Therefore, it is advisable to focus on the model of this shell.

Variants of models for analyzing the behavior of the bottom can be taken as calculation schemes such as round flat shells of rotation, rigidly, pivotally, or elastically fixed to the edge contour. The most preferred option was selected for the hinge support of the edge contour, for which the main results are presented.

The bottom of the container corresponds to the shell of rotation, made up of segments of a sphere in the central part, a torus, and an annular plate on the periphery. Models that take into account the shoulder were also considered. The torus insert smoothly matches the center and edge areas without breaking the normal angle. Parameterization of such a shell is performed, which is associated with the definition of curvatures as functions of the polar radius. It depends on which set of original dimensions is specified.We relied on arrows of the destruction of the sections and the coordinates of the conjugation points on the meridian, Fig. 6.2.2.

The elements of the meridian mate smoothly without breaking, so that the main functions of the resolving systems of static and dynamic problems are continuous at the interface points. Therefore, we can apply the equations written in the reference frame of the trihedron of the middle surface.

So, we consider a shell and, for comparison, a plate of thickness h with an external radius r_d, made of a material with Young's modulus E and Poisson's coefficient ν. When switching to dimensionless values, the characteristic normalizing parameters h_* and R_* are

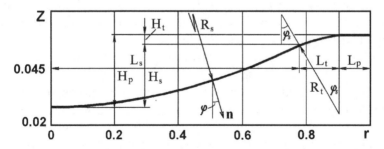

Fig. 6.2.2.   Parameters of the composite shell meridian.

assumed to be equal, respectively, to the thickness of the bottom and its radius: $h_* = h$, $R_* = r_d$. The elastic modulus is normalized by the value $E_* = E$; $\nu_* = \nu$. The thickness h is measured by the displacement components (deflections). The rotation angle is normalized by the parameter $\varepsilon_* = h_*/R_*$.

The arrows of the lines and sections and the coordinates of the interface points on the meridian are assumed to be known. There are the following relations:

$$L_s = k_s r_d, \quad L_p = k_p r_d, \quad L_t = (1 - k_s - k_p), \quad 0 < k_s < 1,$$

$$0 < k_p < 1, \quad 0 < k_s + k_p < 1,$$

$$H_s = k_H H_p, \quad R_s = (H_s^2 + L_s^2)/(2H_s),$$

$$\varphi_s = \arcsin(L_s/R_s), \quad H_s = R_s(1 - \cos\varphi_s),$$

$$H_t = R_t(1 - \cos\varphi_s), \quad H_s + H_t = H_p, \quad R_t = H_p/(1 - \cos\varphi_s) - R_s.$$
$$(6.2.1)$$

Dimensionless parameters of the bottom of the original standard tank were set as follows:

$$h = 1, \quad r_d = 1, \quad E = 1, \quad \nu = 0.3, \quad \varepsilon_* = 3.5 \cdot 10^{-3};$$

$$B = 1, \quad D = 0.0833; \quad k_s = 0.75,$$

$$k_H = 0.8, \quad k_t = 0.1875, \quad k_p = 0.0625,$$

$$R_s = 10.94, \quad \varphi_s = 0.0679, \quad R_t = 2.735,$$

$$H_p = 0.315, \quad H_s = 0.0252, \quad H_t = 0.0063.$$

Parameter $\psi = 1/r$. The hydrostatic pressure is created by a column of liquid with a height of $H_L = 3.065$, which corresponds to the filling coefficient of the tank volume equal to 0.95. In this case, the dimensionless load on the bottom is $q_3 = 3.1 \cdot 10^{-4}$.

The curvature of $k_1(r) = d\varphi(r)/dr$ of the meridian is equal to $1/R_s$ on the spherical section, $-1/R_t$ on the torus, and zero on the plate. The curvature $k_2(r) = \varphi(r)/r$ is equal to $1/R_s$ on a section of the sphere; it decreases linearly from this value to zero on the torus insert and remains zero on the plate.

Thus, the curvature k_1 of the meridian is a piecewise constant alternating function. The curvature of k_2 is continuous. Over the entire radius of the bottom, discontinuous curvature functions were defined as superpositions of piecewise defined functions set using logical operators. The spatial visualization of the bottom shapes is shown

Fig. 6.2.3. Model of the bottom without a shoulder.

Fig. 6.2.4. Model of the bottom with a shoulder.

in Fig. 6.2.3 (without collar) and Fig. 6.2.4 (with collar). In the calculations, the model without a collar was used mainly as a fairly adequate and convenient one.

Variants of bottoms formed by a plate in the center, a conical shell, and an annular plate adjacent to the edge contour are also considered.

6.3. Stress State of the Bottom Loaded with the Weight of the Liquid

In mathematical modeling, the equations of axisymmetric stress–strain state of thin elastic shells of rotation were used (Chapter 8). For reliability and control of the results, two versions of the theories were used: a quadratic-nonlinear theory of V.V. Novozhilov-type based on Kirchhoff and E. Reissner hypotheses with large rotation angles and taking into account the transverse shift, as well as their linearized analogues.

So, we proceed from the equations of the axisymmetric stress–strain state of thin elastic shells of rotation in the framework of the quadratic-nonlinear theory and Kirchhoff hypotheses. Radial axisymmetric deformation is considered, so that all functions depend only on the meridional coordinate α_1. When implementing the algorithm, the equations were written in dimensionless form.

The solving equations are reduced to a system of ordinary differential equations of the sixth order of the canonical form

$$dy_j/d\alpha_1 = A_1(f_j + b_j + f_j^*), \quad j = 1, \ldots, 6, \tag{6.3.1}$$

with the right parts

$$f_1 = \psi(T_{22} - y_1) - k_1 y_2, \quad f_2 = -\psi y_2 + k_1 y_1 + k_2 T_{22},$$

$$f_3 = \psi(M_{22} - y_3) + y_2/\varepsilon_*, \quad f_4 = E_{11} - y_1 - k_1 y_5,$$

$$f_5 = k_1 y_4 - y_6, \quad f_6 = k_{11}/\varepsilon_*;$$

$$b_1 = -q_1, \quad b_3 = -q_3;$$

$$f_1^* = 0, \quad f_2^* = 0, \quad f_3^* = y_1 y_6,$$

$$f_4^* = -0.5\varepsilon_*(y_6)^2, \quad f_5^* = 0, \quad f_6^* = 0. \tag{6.3.2}$$

Here, A_1 is Lame coefficient on the meridian,

$$E_{22} = \psi y_4 + K_2 y_5, \quad K_{22} = \varepsilon_* \psi y_6,$$

$$k_{11} = y_3/D - \nu K_{22}, \quad E_{11} = y_1/B - \nu E_{22},$$

$$T_{22} = B(E_{22} + \nu E_{11}), \quad M_{22} = D(K_{22} + \nu K_{11}). \tag{6.3.3}$$

The main functions in the equations are

$y_1 = T_{11}$ — tension-compression force (in the direction of the meridian);

$y_2 = Q_{11}$ — cutting force;

$y_3 = M_{11}$ — bending moment;

$y_4 = u$ — displacement in the direction of tangent to the meridian;

$y_5 = w$ — displacement along the normal to the shell;

$y_6 = \vartheta_1$ — the rotation angle of the shell element relative to the tangent to the circumferential coordinate line.

In closed form, the right parts (6.3.2) of the system (6.3.1) expressed completely in terms of independent functions yj have the form for an isotropic shell

$$f_1 = \psi[(\nu - 1)y_1 + B(1 - \nu^2)(\psi y_4 + k_2 y_5)] - K_t k_1 y_2,$$

$$f_2 = -\psi y_2 + (k_1 + \nu k_2)y_1 + k_2 B(1 - \nu^2)(\psi y_4 + k_2 y_5) - q_3,$$

$$f_3 = \psi[(\nu - 1)y_3 + D(1 - \nu^2)\varepsilon_1 \psi y_6] + y_2/\varepsilon_* + K_n y_1 y_6,$$

$$f_4 = y_1/B - \psi \nu y_4 - (k_1 + \nu k_2)y_5 - 0.5 K_n \varepsilon_*(y_6)^2,$$

$$f_5 = -y_6 + K_t k_1 y_4, \quad f_6 = y_3/(\varepsilon_* D) - \psi \nu y_6. \tag{6.3.4}$$

For the variant of shallow shells in the right parts of f_1 and f_5 from (6.3.4), you can omit the summands $k_1 y_2$ and $k_1 y_4$, and also consider the current radius r of the polar coordinate system on the

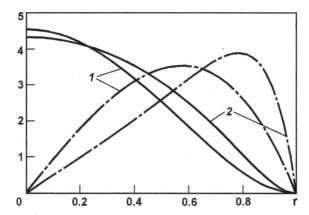

Fig. 6.3.1.   Deflection and angle of rotation of the plate: 1. linear model, 2. nonlinear model.

plane as an independent coordinate ($\alpha_1 = r$). In the ratios (6.3.4), this model corresponds to the value $K_t = 0$. For a non-shallow shell, $K_t = 1$. The values of the K_n switch from 0 to 1 to distinguish linear and nonlinear models, respectively.

The stress–strain state of the tank bottom was analyzed based on linear and nonlinear models, as well as plates of similar thickness and radius.

Figure 6.3.1 shows the deflections w (solid curves) and rotation angles ϑ_1 (dash-dotted lines) for the plate. To place curves on a single field, scaling coefficients are used: for a linear model $0.1w_l$ and $0.05\vartheta_{1l}$ are given, for a nonlinear model, w_n and $0.5\vartheta_{1n}$ are given. Similar curves for the shell are given in Fig. 6.3.2; w_l, $0.5\vartheta_{1l}$, w_n and $0.5\vartheta_{1n}$ are shown.

It can be seen that the behavior of the shell differs significantly from the behavior of the plate. The deflections of the plate exceeds four times the thicknesses in the center. The shell is better adapted to resist distributed pressure and is more rigid. However, the deflections of the shell are comparable to the thickness of the bottom. Therefore, it is necessary to apply nonlinear models for the considered elements.

The stress intensity graphs for the bottom are shown in Fig. 6.3.3. Here, the solid curve corresponds to the stress intensity σ on the middle surface $z = 0$, the dash-dotted line — σ on the outer face

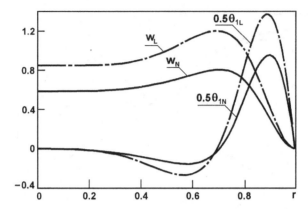

Fig. 6.3.2.   Deflection and angle of rotation of the shell: dash-dotted line — linear model, solid curves — nonlinear model.

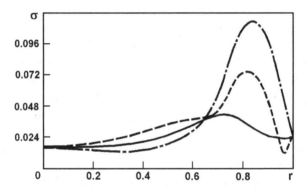

Fig. 6.3.3.   Stress intensity.

surface $z = 0.5$, and the dashed line — σ on the inner surface $z = -0.5$ of the shell.

Figure 6.3.4 shows the meridional and ring generalized internal forces obtained from the nonlinear model. The solid curve is the force T_{11}, the dash-dotted line is the force T_{22}, and the dashed line and point curve are the moments M_{11} and M_{22} respectively, multiplied by a factor of 2.5.

The calculations show that the greatest deflections and rapid variability of the stress–strain state of the bottoms occur in the zone of the torus insert and in the vicinity of the transition line of its interface with the sphere. Here the sign of Gaussian curvature changes. In the

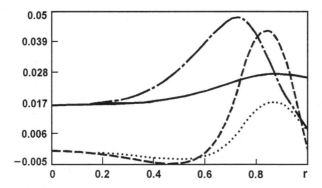

Fig. 6.3.4.   Internal forces and moments.

same zone, there are peak values of stress intensity (Fig. 6.3.3) and a component of the static stress–strain state (Fig. 6.3.4). This type of stress distribution allows us to explain the places of occurrence and types of cracks observed in experiments and in practice.

6.4.   Comparison of Theory and Experiment

The obtained results of theoretical modeling allow us to explain the results of experiments and practical observations carried out. In the laboratory of experimental mechanics of the University of Dortmund, in terms of the location and type of fatigue cracks, the cracks originate in the zone of the highest stress intensity, which is located near the line of change of the sign of the meridian curvature (or Gaussian curvature), Figs. 6.2.2 and 6.3.3. Here, there are also increased bending stresses, and the ratio between the radial and circumferential membrane forces T_{11} and T_{22} changes, Fig. 6.3.4. In the area of approximately 0.8–0.9 of the bottom radius, T_{11} is less than T_{22} $(T_{11} < T_{22})$, and radially oriented cracks are formed here. Moreover, since this ratio is maintained at the peak of stress intensity, this type of crack is more likely. However, closer to the support ring, the ratio between forces changes to the opposite. At the same time, the overall intensity of stress is still high. Therefore, the cracks oriented in the direction of parallels are also likely. At the parallel, where $T_{11} = T_{22}$, cross-shaped cracks are likely. These observed variants are shown in Fig. 6.4.1.

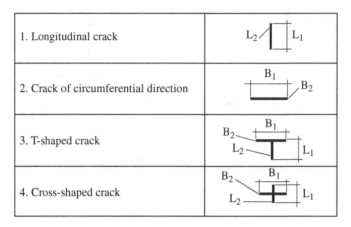

1. Longitudinal crack	L_2 L_1
2. Crack of circumferential direction	B_1 B_2
3. T-shaped crack	B_2 B_1 L_2 L_1
4. Cross-shaped crack	B_2 B_1 L_2 L_1

Fig. 6.4.1.   Types of cracks.

Fig. 6.4.2.   Origin of the crack.

The actual implementation depends on the coordinates of the most significant inhomogeneities of the material, initial imperfections, operational defects, etc., which initiate the appearance of microplastic damages and the origin (Fig. 6.4.2), the subsequent development, and growth of cracks (Fig. 6.4.3).

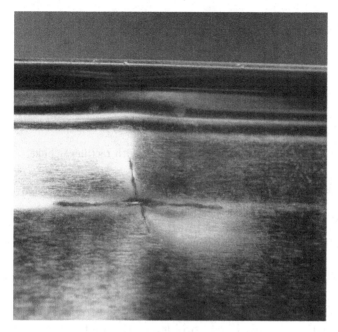

Fig. 6.4.3. Cruciform crack.

6.5. Variants of Shells with a Smooth and Broken Meridian

In Section 6.3, the rotation shell of radius r_d was analyzed, consisting of smoothly conjugated segments of a sphere in the central part, a torus, and an annular plate on the periphery (Fig. 6.2.2). The shell is loaded with hydrostatic pressure created by a column of liquid with a height $H_L = 3.065r_d$. We consider it a shell of the first type.

As a construction of the second type, consider a rotation shell with the same outer radius r_d, composed of a circular plate of radius L_{pc} in the center, a truncated cone, and an annular plate of width L_{pe} in the edge zone. The meridian of such a shell is formed by rectilinear segments connected to the fracture of the generatrix, Fig. 6.5.1. The projection of the generatrix of the cone to the plane of its Foundation marked the height of the cone — Hc. The angle of inclination of the external normal of the forming cone to the axis of symmetry $\beta_c = \text{arctg}(H_p/L_c)$. The shell elements are also made of an isotropic sheet material of thickness h, with Young's modulus E

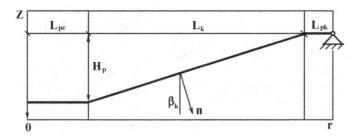

Fig. 6.5.1.   Meridian of the shell with rectilinear links.

and Poisson's ratio ν. The outer contour of the shells is assumed to be hinged.

The shell in question is interpreted as a more technologically advanced type of tank bottom, an alternative to the version with smoothly mated segments.

To model shells of the second type, the equations of Section 8.4 were used. The boundary value problem was solved by targeting with one-way or two-way integration of Cauchy problem. To set initial approximations in nonlinear problems, the method of continuation by the nonlinearity parameter k_N was used. First, the linear problem for $k_N = 0$ was solved, then the transition to the nonlinear problem up to $K_n = 0$ was performed in steps along K_n. Ten iterations were usually sufficient for this purpose.

The bottom is the most loaded element of the container. As shown by the research presented in Chapters 4 and 5, its stress state determines the resource of the entire structure. At the same time, knowledge of the static stress–strain state is sufficient to evaluate the resource.

The stress–strain state of the initial variant of the bottom with smooth bends of the meridian was studied in Section 6.3. Below is an analysis and comparison of eight variants of the two types of bottoms loaded with the weight of the liquid when filling the container at 95%.

When switching to dimensionless values, the characteristic normalizing parameters h_* and R_* were assumed to be equal, respectively, to the thickness of the bottom and its radius: $h_* = h$ and $R_* = r_d$. The elastic modulus is normalized by the value $E_* = E$; $\nu_* = \nu$. The displacement components are based on the thickness.

Table 6.5.1.   Shell parameters of the first type.

No	K_H	L_S	L_T	L_P	$Max(\sigma)$
1.1	0.8	0.75	0.1875	0.0625	0.124
1.2	0.8	0.78	0.195	0.025	0.113
1.3	0.947	0.9	0.05	0.05	0.122
1.4	0.8	0.8	0.2	0	0.104
1.5	1.0	1.0	0	0	0.030

Table 6.5.2.   Parameters of the second type of shell.

No	H_p	L_{pc}	L_c	L_{pe}	$Max(\sigma)$
2.1	0.0315	0.175	0.7	0.125	0.142
2.2	0.0315	0.175	0.7625	0.0625	0.109
2.3	0.0315	0.175	0.825	0	0.049

The angle of rotation is normalized by the parameter $\varepsilon_* = h_*/R_*$ and the intensity of stress, by the value $E\varepsilon_*$.

In the dimensionless form, the arrow for the H_p of the shells was set to the same value, equal to 0.0315. Pressure at the edge plate level $q_3 = 0.0031$. Due to the flatness of the bottom, it is almost constant in radius. Other general parameters are $h = 1$, $r_d = 1$, $E = 1$, $\nu = 0.3$, $\varepsilon_* = 3.5 \cdot 10^{-3}$, $B = 1$, $D = 0.0833$. The variable parameters corresponding to the sections as a fraction of the outer radius of the shell are summarized in Table 6.5.1 for the first type of bottom and in Table 6.5.2 for the second type of bottom.

Let's consider some results for option No. 2.1. in the case of shells with meridian breaks, as for a smooth shell it is necessary to rely on the solution of a nonlinear problem.

Figure 6.5.2 shows normal deflections for the nonlinear (solid curve) and linear (dash-dotted line) problems for option No. 2.1. The dashed curve corresponds to the smooth shell No. 1.1 from the solution of the nonlinear problem. The deflection levels for the two types of shells are close.

The stress intensities on the middle and front surfaces of the shell No. 2.1 are shown in Fig. 6.5.3. The solid curve corresponds to $\sigma(0)$

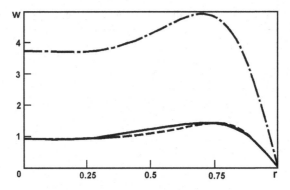

Fig. 6.5.2.   Graphics of normal deflections.

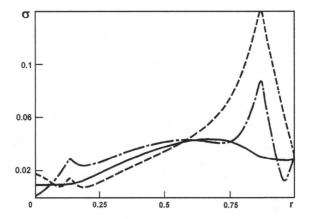

Fig. 6.5.3.   The stress intensity for the shell No. 2.1.

on the middle surface, the dashed line corresponds to $\sigma(0.5h)$ on the outer (external) surface from the positive direction of the normal, and the dash-dotted line corresponds to the intensity of $\sigma(-0.5h)$ on the inner surface.

The maximum stress intensity is achieved at the junction of the edge plate with the cone on the outer face surface. The shape of dependencies is clearly affected by the shell geometry.

Figure 6.5.4 compares the stress intensity on the outer face of the smooth shell No. 1.1 (dash-dotted line) and shell No. 2.1 (solid curve).

The maximum intensity for a non-smooth shell is slightly higher ($\sim 15\%$), but not radically.

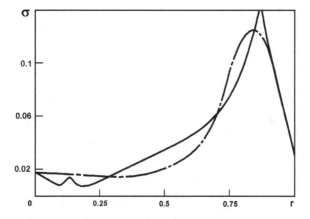

Fig. 6.5.4. Comparison of shell stress intensity two types.

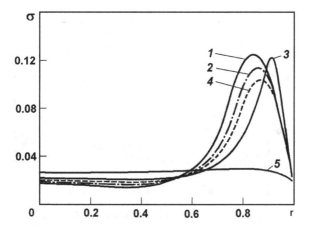

Fig. 6.5.5. Stress intensity of the first type shells.

Calculations of the stress intensity σ for other shell variants show that the maximum is also reached on the outer face surface. The dimensionless maximum values are given in Tables 6.5.1 and 6.5.2 in the rightmost columns. The radius distributions σ on this surface are shown in Fig. 6.5.5 for shells of the first type and Fig. 6.5.6 for shells of the second type.

In the presence of an edge plate, there is a significant surge of stress in the vicinity of the inner radius of the ring plate. Reducing the width of the edge plate reduces the stress concentration. The best results are given by variants without a flat edge zone (for example,

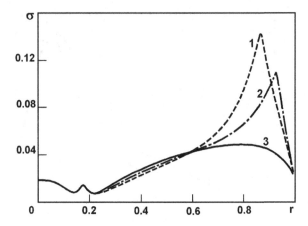

Fig. 6.5.6.   Stress Intensity of the second type shells.

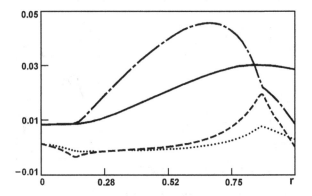

Fig. 6.5.7.   Internal forces and moments of the shell No. 2.1.

curve 3 in Fig. 6.5.6). In the minimal tension and good uniformity of
the stress-strain state, there is a bottom formed by a single spherical
segment (curve 5 in Fig. 6.5.5).

Thus, for shells of the first type, the transition zones from the
spherical shell through the torus insert to the plate are stress intensity concentrators. Stress intensity peaks also occur at meridian
breaks in the vicinity of the boundary zone for shells of the second type. Here, as for the original shell of the first type, there is an
annular line of inversion of the membrane forces.

Figure 6.5.7 shows the meridional and annular internal forces for variant No. 2.1 obtained from the nonlinear model. The results are presented for dimensionless quantities. Solid curve for force T_{11}, dash-dotted line — force T_{22}, dashed line — moment M_{11}, point curve — moment M_{22}.

In the vicinity of the inversion line of internal forces, annular, radial, and cruciform cracks may also occur. Therefore, it is advisable to avoid sudden changes in the shape of the meridian when approaching the edge ring. The presence of a flat ring zone is clearly undesirable. Its possible removal should not significantly complicate the technology of rolling the edge, since the bottom is a very flat shell.

6.6.   Shell with an Annular Groove

The bottom designs are quite common, in which grooves of various configurations are stamped to increase the bending stiffness, Fig. 6.6.1. Among them there is a variant with ring grooves (bottom 4–1).

Consider the shell of type No. 2.1, which has an annular groove at the junction of the cone and the edge plate, Fig. 6.6.2.

Figure 6.6.3 shows the deflections of the shell with a groove (solid curve) and without it (dashed line), obtained by solving a nonlinear

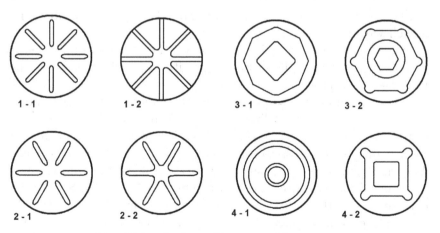

Fig. 6.6.1.   Designs of bottoms with grooves.

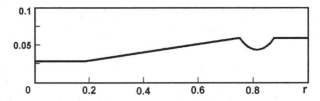

Fig. 6.6.2.   Meridian of the shell type No. 2.1 with an annular groove.

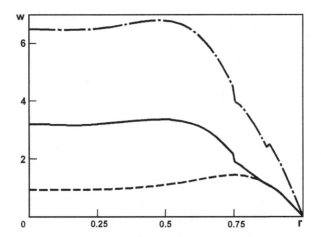

Fig. 6.6.3.   Deflections of the shell with and without a groove.

problem. The dash-dotted curve corresponds to a linear problem for a shell with a groove.

The shell with a groove is more pliable: the deflections in its center increase three times. The stress intensity shown in Fig. 6.6.4 also increases almost twice in this case.

Here, as before, the solid curve corresponds to $\sigma(0)$ on the middle surface, the dash-dotted line corresponds to $\sigma(0.5h)$ on the outer (outer) surface from the positive direction of the normal, and the dashed line corresponds to intensity $\sigma(-0.5h)$ on the inner surface. The maximum stress intensity is at the bottom of the groove.

Internal forces and moments are shown in Fig. 6.6.5. Solid curve corresponds to the stress T_{11}, the dash-dotted line stress to T_{22}, the dashed line to the moment M_{11}, and point curve to point M_{22}. Peaks are observed at the bottom and edges of the groove.

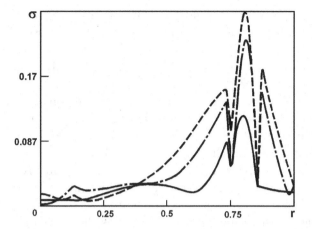

Fig. 6.6.4.   Stress intensity of the shell with an annular groove.

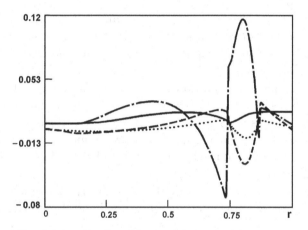

Fig. 6.6.5.   Graphs of internal forces and moments.

Thus, the grooves are a factor that increase the tension and heterogeneity of the shell state. This also reduces the life of the shell, bringing the cyclic strength to low-cycle fatigue.

6.7.   Changing the Shape of the Tank Bottom

This section deals with the problem of changing the shape of the shell, modeling the bottom of the axisymmetric tank for the transportation of liquid cargo. The original shell of composite geometry is transferred in the form of a spherical segment. The purpose of shaping is to reduce the level of stress intensity. Mathematical modeling is performed on the basis of geometrically and physically nonlinear equations using the semi-inverse method.

There is a wide class of structures that are containers for transportation of liquid cargo [48, 49]. During transportation, the body of the thin-walled tank experiences both static and dynamic loads due to fluctuations in the vehicle. Multiple variable loads can cause fatigue cracks and leakages. Given the specific nature of the substances transported, this can cause environmental and economic damage. Therefore, it is advisable to analyze the causes of destruction of widely used tanks in practice, assess the shortcomings of some structures, search for technology to eliminate them, and compare alternative options.

Studies show that static stresses caused by the weight of the load have a significant impact on the resource of structures. The widespread and typical capacities include two-hundred-liter barrels, Fig. 6.2.1. The most loaded and weak element of their design is the bottom, Fig. 6.7.1. A common variant of the bottom of the composite geometry is one in which there is a section of the annular plate (edge), a shallow spherical dome (center), and a torus insert between them. The Gaussian curvature $K = k_1 k_2$ of such a shell changes the sign on the interface line of the sphere and the torus; k_1, k_2 are the main curvatures.

The shell is subjected to radial axisymmetric deformation, so that all functions depend only on the meridional coordinate α_1. The study

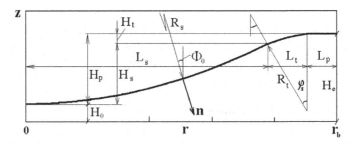

Fig. 6.7.1.   The sketch of the tank and the original shape of the bottom.

of the shell of this form shows that its stress state is not optimal. In the zone of the torus insert, there is a surge of stress intensity, which leads to the appearance and development of fatigue cracks during operation, leading to destruction.

At the same time, the bottom in the form of a shallow spherical shell without sections of other geometry gives a smoothed stress state, which increases the resource (Section 6.3).

Containers with composite geometry bottoms are stamped by hundreds of thousands in mass production. Due to commercial reasons, the manufacturer does not benefit from a very long service life of the products. The carrier is interested in the opposite, both in durability and reliability. It is responsible for the environmental consequences of leakage of harmful liquids into the environment.

Is it possible to improve their properties by changing the shape of the bottoms if the carrier has a large number of such tanks? We set the following problem: from the shell of the composite geometry having an alternating Gaussian curvature K, we obtain a spherical (or close to it) shell ($K > 0$), which gives the best characteristics of the stress state.

For mathematical modeling of shell forming, geometrically and physically nonlinear equations are used, which allow large displacements and angles of rotation, change of the metric of the middle surface, and compression of the material normal. When you build equations, use exponential approximation diagrams of hardening material and deformational physical relations with logarithmic relative elongations. The semi-inverse method, considered in Chapter 5, is adapted to the specifics of the problem related to the geometry of the original shell [54].

Kinematic relations are similar to those presented in the previous section and have the form

$$e_1 = (\varepsilon_1 + \zeta\kappa_1), \quad e_2 = (\varepsilon_2 + \zeta\kappa_2),$$

$$\varepsilon_1 = (w'\sin\Phi + u'\cos\Phi)/\alpha_o + \cos(\Phi - \Phi_o) - 1,$$

$$\varepsilon_2 = u/r_o, \quad \kappa_1 = \Phi_o'/\alpha_o - (1 + \varepsilon_3)K_1,$$

$$\kappa_2 = (\sin\Phi_o)/r_o - (1 + \varepsilon_3)K_2,$$

$$K_1 = \Phi'/\alpha_o, \quad K_2 = (\sin\Phi)/r_o; \tag{6.7.1}$$

$$\gamma_1 = \gamma_o/(1 + \varepsilon_1),$$

$$\gamma_o = (w'\cos\Phi - u'\sin\Phi)/\alpha_o - \sin(\Phi - \Phi_o); \qquad (6.7.2)$$

$$u' = \alpha_o[\varepsilon_1\cos\Phi - \gamma_o\sin\Phi + (\cos\Phi - \cos\Phi_o)],$$

$$w' = \alpha_o[\varepsilon_1\sin\Phi + \gamma_o\cos\Phi + (\sin\Phi - \sin\Phi_o)]. \qquad (6.7.3)$$

Here, all functions depend on the curvilinear coordinate $\alpha_1 = \xi$. The prime denotes the derivative with respect α_1; α_o is the initial shell Lame ratio; Φ_o is the angle between the normal to the original shell and the axis z; Φ is the angle between the normal to the deformed shell and the axis z; $\delta_j = 1 + \varepsilon_j$, $j = 1, 2, 3$. The incompressibility condition is assumed to be satisfied: $\delta_1\delta_2\delta_3 = 1$, $\delta_3 = (\delta_1\delta_2)^{-1}$. Equilibrium equations are as follows:

$$(r_o\bar{V}^o)' + \alpha_o r_o p_w^o = 0; \qquad (6.7.4)$$

$$(r_o\bar{H}^o)' - \alpha_o\bar{N}_2^o + \alpha_o r_o p_u^o = 0; \qquad (6.7.5)$$

$$(r_o M_1^o)' - \alpha_o M_2^o\cos\Phi - \alpha_o r_o\delta_1 Q^o = 0. \qquad (6.7.6)$$

$$p_w^o = \delta_1\delta_2 p_w, \quad p_u^o = \delta_1\delta_2 p_u. \qquad (6.7.7)$$

Internal effort $\bar{V}^o$ and $\bar{H}^o$ are oriented, respectively, vertically (along the axis of symmetry) and horizontally (along the radius of the cylindrical coordinate system) and Q^o is the cutting force:

$$\bar{V}^o = \bar{N}_1^o\sin\Phi + Q^o\cos\Phi,$$

$$\bar{H}^o = \bar{N}_1^o\cos\Phi - Q^o\sin\Phi. \qquad (6.7.8)$$

Values $\bar{N}_1^o$, $\bar{N}_2^o$ are generalized efforts associated with internal tangential efforts N_1^o, N_2^o and moments M_1^o, M_2^o.

$$\bar{N}_1^o = N_1^o + \delta_1^{-1}(K_1 M_1^o + K_2 M_2^o),$$

$$\bar{N}_2^o = N_2^o + \delta_2^{-1}(K_1 M_1^o + K_2 M_2^o),$$

$$N_1^o = \delta_2 N_1, \quad N_2^o = \delta_1 N_2, \quad Q^o = \delta_2 Q,$$

$$M_1^o = \delta_2 M_1, \quad M_2^o = \delta_1 M_2. \qquad (6.7.9)$$

Internal pressure is considered as a load. Given the positive direction of the normal inside the shell, we have

$$p_w = -p\cos\Phi, \quad p_u = p\sin\Phi. \qquad (6.7.10)$$

Defining relations of the Davis–Nadai type are accepted in the form

$$N_1^o = \bar{B}_1(\bar{\varepsilon}_1 + 0.5\bar{\varepsilon}_2), \quad N_2^o = \bar{B}_2(\bar{\varepsilon}_2 + 0.5\bar{\varepsilon}_1),$$

$$M_1^o = \bar{D}_1(\bar{\kappa}_1 + 0.5\bar{\kappa}_2), \quad M_2^o = \bar{D}_2(\bar{\kappa}_2 + 0.5\bar{\kappa}_1); \quad (6.7.11)$$

$$\bar{\varepsilon}_1 = \ln(\delta_1), \quad \bar{\varepsilon}_2 = \ln(\delta_2), \quad \bar{\kappa}_1 = \kappa_1/\delta_1, \bar{\kappa}_2 = \kappa_2/\delta_2; \quad (6.7.12)$$

$$\bar{B}_1 = \bar{B}/\delta_1, \quad \bar{B}_2 = \bar{B}/\delta_2, \quad \bar{D}_1 = \delta_3^2\delta_2\bar{D}, \quad \bar{D}_2 = \delta_3^2\delta_1\bar{D},$$

$$\bar{B} = (4/3)\Lambda(\bar{\varepsilon})h_o, \quad \bar{D} = (1/9)\Lambda(\bar{\varepsilon})h_o^3. \quad (6.7.13)$$

Here, $\Lambda(\bar{\varepsilon}) = C\bar{\varepsilon}^{\eta-1}$ is the secant modulus of the graph of the ductility of the material, where

$$\bar{\varepsilon} = (2/\sqrt{3})\sqrt{\bar{\varepsilon}_1^2 + \bar{\varepsilon}_1\bar{\varepsilon}_2 + \bar{\varepsilon}_2^2}$$

is the intensity of the logarithmic strain.

The material diagram can be identified by characteristic values, which are usually given in reference books. It is the modulus of elasticity E, the limit of elasticity σ_{02}, the limit of elastic deformation $\varepsilon_{02} = \sigma_{02}/E$, the limit stress intensity σ_l, and the limit intensity of the logarithmic strain $\bar{\varepsilon}_l$. According to these data, the parameters of the power approximation of the diagram of nonlinear properties of the material are determined.

$$\sigma(\bar{\varepsilon}) = C\bar{\varepsilon}^{\eta},$$

where

$$\eta = [\ln(\sigma_l) - \ln(\sigma_{02})/[\ln(\bar{\varepsilon}_l) - \ln(\varepsilon_{02})], \quad C = \sigma_l/\bar{\varepsilon}_l^{\eta}.$$

So, for the considered material, capacity (steel) characteristics were set as follows: $E = 0.21 \cdot 10^6\,MPa$, $\sigma_{02} = 210\,MPa$, $\sigma_l = 380\,MPa$, $\varepsilon_{02} = 0.001$, $\bar{\varepsilon}_l = 0.25$, $C = 44.1\,MPa$, and $\eta = 0.107$. The secant modulus was set equal to E for $\bar{\varepsilon} < \varepsilon_{02}$ and $\Lambda(\bar{\varepsilon})$ when $\bar{\varepsilon} \geq \varepsilon_{02}$.

Let's integrate equations (6.7.4) and (6.7.5). Given (6.7.8), we obtain the following:

$$r_o(\bar{N}_1^o \sin\Phi + Q^o \cos\Phi) = p \cdot I_{11} + C_V, \quad (6.7.14)$$

$$r_o(\bar{N}_1^o \cos\Phi - Q^o \sin\Phi) = I_{21} - p \cdot I_{22} + C_H, \quad (6.7.15)$$

where C_V and C_H are integration constants associated with vertical and horizontal forces at one end of the integration interval. In this case, we will integrate from the edge of the shell to the center to set the forces on the outer contour r_b. Then I_{11}, I_{21}, I_{22} depending on the current coordinate are written as

$$I_{11} = -\int_{\xi}^{r_b} \alpha_o r_o \delta_1 \delta_2 \cos \Phi d\xi, \quad I_{21} = -\int_{\xi}^{r_b} \alpha_o \bar{N}_2^o r_o d\xi,$$

$$I_{22} = -\int_{\xi}^{r_b} \alpha_o r_o \delta_1 \delta_2 \sin \Phi d\xi. \tag{6.7.16}$$

From the relations (6.7.14) and (6.7.15), we express Q^o. To do this, multiply the equation (6.7.14) by $\cos \Phi$ and subtract from it the equation (6.7.15) multiplied by $\sin \Phi$. We have

$$Q^o = r_o^{-1}[(p \cdot I_{11} + C_V) \cos \Phi$$
$$- (I_{21} - p \cdot I_{22} + C_H) \sin \Phi]. \tag{6.7.17}$$

Substitute the expression (2.17) for Q^o in equation (2.6) and integrate it. We get

$$r_o M_1^o = I_{31} + p \cdot (J_{11} + J_{22}) + C_V J_{31}$$
$$- J_{32} - C_H J_{33} + C_M, \tag{6.7.18}$$

where

$$J_{11} = -\int_{\xi}^{r_b} \alpha_o \delta_1 I_{11} \cos \Phi d\xi, \quad J_{22} = -\int_{\xi}^{r_b} \alpha_o \delta_1 I_{22} \sin \Phi d\xi,$$

$$J_{31} = -\int_{\xi}^{r_b} \alpha_o \delta_1 \cos \Phi d\xi, \quad J_{32} = -\int_{\xi}^{r_b} \alpha_o \delta_1 I_{21} \sin \Phi d\xi,$$

$$J_{33} = -\int_{\xi}^{r_b} \alpha_o \delta_1 \sin \Phi d\xi. \tag{6.7.19}$$

C_M is the integration constant associated with the moment reaction on the circuit.

The pressure p does not depend on ξ and is taken out of the integral sign. From (6.7.18), it follows that

$$p = \frac{r_o M_1^o + J_{32} - J_{31} - C_M - C_V J_{31} + C_H J_{33}}{J_{11} + J_{22}}. \tag{6.7.20}$$

The output of the expression (6.7.20) to a constant is controlled by the C_V, C_H, C_M parameters and the geometry of the new shell shape.

The shell material is ordinary steel type 3. The initial shape of the shell is composed of segments of the sphere (central part), torus, and plate (boundary zone), Fig. 6.7.1. The torus inserts smoothly, without breaking the angle of the n normal slope Φ_o to the axis of symmetry and matches the central and edge areas. The shell has a thickness $h = 1\,mm$ and outer radius $r_b = 282.75\,mm$. Other parameters were set as follows (in mm):

$$L_s = 212, \quad L_t = 53, \quad L_p = 17.7, \quad H_t = 1.8, \quad H_e = 17, \quad H_o = 8,$$

$$H_p = H_e - H_o = 9, \quad H_s = 7.2,$$

$$\varphi_s = \arcsin(L_s/R_s) = 0.067878, \quad H_s = R_s(1 - \cos\varphi_s),$$

$$R_s = [(H_s)^2 + (L_s)^2]/(2H_s) = 3126.$$

The curvature of the meridian $k_1 = d\Phi_o/dr$ is a piecewise constant alternating function equal to $1/R_s$ in the spherical region, $-1/R_t$ in the torus, and zero in the flat region. The second main curvature $k_2 = (\sin\Phi_o)/r_o$ is continuous, equal to $1/R_s$ in the area of the sphere, and decreases linearly from this value to zero on the torus insert when approaching the plate and remains zero on the plate.

We take as an independent coordinate the length of the arc of the original shell, the lengths of the sections are

$$s_1 = R_s\varphi_s, \quad s_2 = s_1 + R_t\varphi_s, \quad s_3 = s_2 + L_p.$$

The equations of a middle surface on sites (vertical coordinate) have the following form:

in the interval of the sphere $0 < s \le s_1$

$$Z_{os}(s) = H_o + R_s[1 - \cos(s/R_s)];$$

on the torus interval $s_1 < s \le s_2$

$$Z_{ot}(s) = H_o + R_s + R_t\{\cos[\varphi_s - (s - s_1)/R_t] - \cos(\varphi_S)\};$$

in the interval of the plate $s_2 < s \le s_3$

$$Z_{op}(s) = H_o + H_p.$$

Normal angles of inclination, respectively, are

$$\Phi_{os}(s) = s/R_S, \quad \Phi_{ot}(s) = \Phi_{os}(s_1) - (s - s_1)/R_t, \quad \Phi_{op}(s) = 0.$$

Polar radius is

$$r_{os}(s) = r_s \cdot \sin(\Phi_{os}(s)), \quad r_{ot}(s) = s_1 + L_t - R_S \cdot \sin(\Phi_{ot}(s)),$$

$$r_{op}(s) = s.$$

Superpositions of these functions on the whole site are denoted, respectively, as $Z_o(s)$, $\Phi_o(s), r_o(s)$. When counting in an arc, $\alpha_o(s) = \alpha_o = 1$.

The expected shape of the shell after deformation is given as a segment of the sphere, the meridian of which passes through the points with coordinates $(0, H_o)$ и (r_b, H_e).

$$Z_d(r) = R_d + H_o - \sqrt{R_d^2 - r^2}, \tag{6.7.21}$$

where R_d is the radius of the segment and H_o is the clearance height. In this instance, $R_d = 4.45\,m$.

The meridian of the shell has a curved shape. Therefore, it can be assumed that its deformation will be predominantly flexural. That is, the lengthening of the meridian will be small, $\varepsilon_1 \ll 1$. This simplifies the application of the method compared to the options of drawing the dome from the plate and inflating the cylinder, where all elongations are of the same order. Therefore, the equation $Z_s(s) = Z_d(r_o(s))$ corresponds to a record in Lagrangian coordinates. The total length of the Meridian of the S_d shell in the deformed state is approximately equal to the original length S_o. The residual $\varepsilon_1 = (S_d - S_o)/S_o$ can be taken as the average elongation.

The condition of compatibility of deformations follows from the relations (6.7.1), (6.7.2) as

$$\alpha_o(\varepsilon_1 \cos \Phi - \gamma_o \sin \Phi + \cos \Phi - \cos \Phi_o) - (r_o \varepsilon_2)' = 0 \tag{6.7.22}$$

and the formula

$$w' = \alpha_o(\varepsilon_1 \sin \Phi + \gamma_o \cos \Phi + \sin \Phi - \sin \Phi_o). \tag{6.7.23}$$

Since $r_0 \varepsilon_2 = u$, $\gamma_o \approx 0$, then from (6.7.22) and (6.7.23) we have

$$u' = \alpha_o[(1 + \varepsilon_1) \cos \Phi - \cos \Phi_o)], \tag{6.7.24}$$

$$w' = \alpha_o[(1 + \varepsilon_1) \sin \Phi - \sin \Phi_o)]. \tag{6.7.25}$$

Since the original and deformed shell shapes are specified, these relations (6.7.24), (6.7.25) are integrated

$$u(s) = \int_{r_b}^{s} \alpha_o[(1 + \varepsilon_1)\cos(\Phi(s)) - \cos(\Phi_o(s))]ds, \quad (6.7.26)$$

$$w(s) = \int_{r_b}^{s} \alpha_o[(1 + \varepsilon_1)\sin(\Phi(s)) - \sin(\Phi_o(s))]ds. \quad (6.7.27)$$

The edges of the shell are not shifted, so the integration constants are zero. The direction of integration does not matter.

The distribution of vertical displacements can also be obtained from the difference between the equations of finite and initial forms $w(s) = Z_s(s) - Z_o(s)$, which is an additional test. The graph $w(s)$ is shown in Fig. 6.7.2.

Knowing the radial displacement (6.7.26), we obtain the elongation of the annular areas $\varepsilon_2(s) = u(s)/r_o(s)$, and from the condition of incompressibility $(1+\varepsilon_1)(1+\varepsilon_2(s))(1+\varepsilon_3(s)) = 1$, we find compression. $\varepsilon_3(s)$. In this case, the stress intensity $\varepsilon(s) = (2/\sqrt{3})\varepsilon_2(s)$ and the secant module are calculated as a function of s: $\Lambda_s(s) = \Lambda(\varepsilon(s))$.

The change in the initial shape in the spherical segment corresponds to a pressure $p = 250\,kPa$. The most active plastic deformations capture the torus insert and its surroundings. The new form has a significantly lower stress intensity under elastic loading by the weight of the liquid (Fig. 6.7.3, curve 2) compared to the prototype (Fig. 6.7.3, curve 1). The bottom formed by one spherical segment is in minimal tension and good uniformity of state.

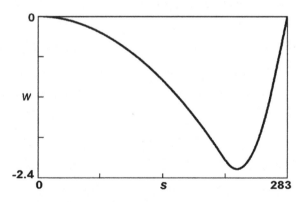

Fig. 6.7.2. Graph $w(s)$ — vertical.

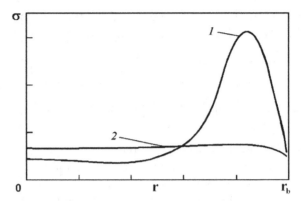

Fig. 6.7.3.   Intensity of stresses in displacement the original (1) and the resulting shells (2).

Let's compare the forming pressure with the test pressures. The tanks are tested for air tightness at a pressure of 20–$30\,kPa$. Hydraulic strength tests are performed by pumping water at a pressure of $100\,kPa$ (up to 250 at large metal thicknesses). Experiments on the limit load show that the destruction during air injection occurs at the level of $\approx 300\,kPa$. In this case, the cylindrical part can withstand this load in the elastic zone.

Destruction occurs by knocking out one of the bottoms. Therefore, the technology of changing the form should be worked out taking into account security. To maintain clearance, the tank must be clamped between the rigid planes on the bottoms. If the connection of the bottoms and shells is made by seaming technology, it is advisable to strengthen the connection using clamps.

Chapter 7

Jack Systems in Construction

This section presents the methods and technologies of alignment of buildings and structures that have received uneven rainfall. The characteristics of the used piston jacks and flat jacks in the form of toroidal shells are given. Aspects of the lifting and leveling process control (on the basis of [50]) are considered.

Uneven deformation of the ground base of buildings under construction and operation and the associated consequences are a serious problem. For example, in Russia, the CIS countries, Poland, and Georgia, hundreds and thousands of objects are in disrepair, 90% of which are residential buildings. A significant number of residential buildings and monuments need to restore operational reliability in Germany, France, Italy, Czech Republic, Brazil, the USA, and other countries. At the same time, we are not talking about dilapidated, has already served its term, buildings and good-quality modern facilities that have received significant uneven precipitation (rolls) during construction or during operation. These facilities can be successfully operated for many years after the removal of uneven sediments (rolls).

Factors of uneven precipitation of construction projects can be divided into the following groups: (1) construction errors; and (2) operation errors, which are most common at the moment. Breakthroughs of water-carrying communications, lack of storm sewers, and poor condition of the blind area lead to intense local soaking of the base soil and, as a result, to uneven deformations.

A number of methods have been developed and applied in the practice of artificial regulation of the position of buildings in space. These include fuming, leveling sandboxes, adjustable soaking of the base, etc. These methods have a number of drawbacks in technological solutions for their application. This is the high cost of equipment, uncontrolled alignment process of the building, etc. Therefore, a wide practical application has been found for a method of alignment with the help of jacks. For these purposes, both piston and flat jacks are used, working on the principle of bellows.

7.1. The Advantages of Flat Jacks in Front of the Piston

Piston jacks are mainly used to adjust low-rise buildings (up to five floors). The great mass of these jacks and big working pressure, of the order of $70\,MPa$, is almost unacceptable for geometric correction of high-rise buildings (16 floors and up).

Piston jacks are divided into two types: (1) piston jacks with a safety nut that allows you to fix the load in the raised position for a long time; (2) piston jacks with ring pads, which are laid in pairs under the rod to insure the load when it is lifted.

Foreign analogs of these types of jacks differ only in design features. In the power units of firms "Hybritech" (Dortmund, Germany) and "DMT" (Essen, Germany), modular systems for lifting and leveling used piston jacks with safety nut, Fig. 7.1.1. The design of the power unit of the firm "Startech" consists of a jack equipped with flat U-shaped inserts, Fig. 7.1.2, and a separately located control unit.

This system has the same command and control elements as discussed above. The speed of movement of the jacks in these systems are regulated by changing the engine speed in each module, which leads to an increase or decrease in the flow of the working fluid entering the cavity of each jack. The system continuously monitors the vertical movement of the jack piston. To ensure the safety of the lifting work of the building, the installation of restrictive liners is carried out in the lower support part of the jack during the lifting process. The lifting principle is step-by-step, while the difference in stroke between two adjacent jacks should not exceed a lifting step of

Fig. 7.1.1. Piston jack with safety nut (firms "Hybritech", "DMT", Germany).

Fig. 7.1.2. Piston jack with U-shaped inserts (firm "Startech", Germany).

$0.5\,mm$. To lift the building to a height of $500\,mm$, the system must work out 1000 microcycles.

The main disadvantages of piston jacks with a load capacity of more than $2000\,kN$ include the following: (a) the large weight of the jack ($95\,kg$ or more), which makes it difficult to work with them in cramped conditions of the basement part of the building; (b) the working pressure of the piston jacks is 70 MPa, and the small diameter of the support part creates contact stresses, in some cases exceeding the specified strength of the concrete structures of the basement part. Despite the shortcomings of systems with piston jacks, they have proven themselves when lifting low-loaded structures and buildings up to 5 floors.

The idea of using the toroidal shell as a jack was proposed by P. Fesin in 1938. The research under the leadership of V. P. Shumovsky conducted in NIISK (Kiev) in 1976–1978 allowed to use the design of the jack for lifting and leveling of buildings and to create an electro-hydraulic system for geometric adjustment of buildings in space. The supply of working fluid to the hydraulic line was carried out from one pumping station. The gidromagistral was located on the inner perimeter of the basement and fastened with brackets on the walls of the raised part of the building. The control was carried out from the console, on which the control equipment of all units of the system, control of the alignment process, and the state of the most critical elements of the hydraulic drive, was mounted. The system was located in two containers, which were transported by road. The weight of only one high-pressure station was 6 tons, and the power of the system was 25 KW.

Flat jacks, made in the form of torus shells, coupled with the plates, have a number of advantages such as a large area of contact with the support surface, low specific fluid pressure in the cavity, high load capacity, good damping and damping properties, reduced weight, and simplicity of construction. A fairly low pressure in the working cavity of the flat jack (12 MPa) with a nominal load capacity of 2000 kN allows you to include in the jack units commercially available automation elements, which makes it possible to automate the process of controlling the rise of the house.

Extensive research and active use of flat jacks are carried out by NPO "Interbiotech" (Russia, Rostov-on-don). Since 1993, the firm studied the mechanics of the working of flat jacks of the toroidal type, and the control systems for the geometric adjustment of buildings in space were improved. Significant practical experience of their application has been accumulated.

Flat jack toroidal shape is a structure in the form of a thin-walled tank of two metal "plates", the outer edges of which are connected by a shell in the form of a torus, Fig. 7.1.3. The system is

Fig. 7.1.3. Flat jacks as the shells of special form.

Fig. 7.1.4. Components of flat jacks.

symmetrical with respect to the middle plane. The technology of production of the jack consists of preliminary cold stamping of "plates", Fig. 7.1.4, with the subsequent welding among themselves on a ring of rigidity.

To increase the height of lifting, jacks are usually folded into packages. The torus allows you to move flat membranes under the action of injection of the liquid (Figs. 7.1.5 and 7.1.6). As a rule,

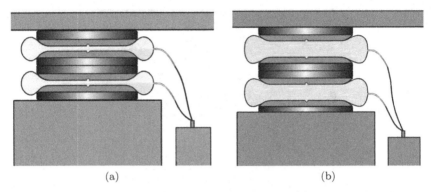

(a)   (b)

Fig. 7.1.5.   Package of two flat jacks in the initial stage (a) and after downloading high pressure oil (b).

Fig. 7.1.6.   Power assembly of flat jacks.

in most cases, the rise and alignment of buildings is carried out without increasing contact stresses. This is provided by the many jacks located in the basement of the building, the height of their installation in relation to the base of the foundation, as well as technological methods of management of jack groups in the process of lifting the building.

There were problems of valve control in the operation of single-flow systems. They are related to the fact that external loads from

Fig. 7.1.7.   Control system of high-pressure multithreaded station.

structural elements of a building are different in magnitude. Also, to arrange the jacks so that the external forces and efforts on the jacks were equal is impossible. The difference in the values of external loads and forces developed on the jacks is 20–$30\,kN$, which causes the operator to constantly turn on or off groups of jacks. It is quite difficult to manage such a system when a section of the building structure is ready later or ahead of other parts of the building.

A fundamentally new system was created (Fig. 7.1.7) in 2006, which uses multithreaded high-pressure stations (2, 4, 6, etc.), depending on the task Stations allow, by changing the pressure of the working fluid in the flows, to bring the force values in the jacks as close as possible to external loads. The system based on multi-threaded high-pressure station allows you to create of various pressure $P_1 \neq P_2 \neq P_3 \neq P_4 \neq \cdots \neq P_n$ in each group of jacks and change their size in the process. The management of the new system is not difficult. If a section of the building structure is built later or ahead of other groups of jacks, it is easily eliminated by the discharge or increase in pressure in the hydraulic line. The possibilities inherent in the system are extremely important when adjusting the position of the building in space, since the value of external loads changes during the lifting process.

The change in the external load from the vertical movements of the building has not been studied. Theoretical and experimental studies in this direction are conducted by employees of the company. By now, over 25 years of work in the association "Interbiotech" has led to more than 10 technologies of alignment of buildings being developed As experience has shown, the technological aspect is the most important when lifting and leveling buildings using jacks. Along with the use of modern hydraulic components and flat jacks of toroidal type, developed in the company, microprocessor control systems over movements and many other technical innovations, the basis for successful and safe work on the rise of buildings is the technology of management of the process of lifting the building.

7.2. Variations of Flat Jacks Geometry

Buildings at maintenance can expose to the exposures, in the form of various sorts of non-uniform strains or tilts. There is production engineering of ascending gradient and leveling of multistory buildings by means of steel shells of the closed volume termed as flat jacks [60]. Flat jacks represents two round plates which are, at the outer edge, connected by torus shell. In interior volume from hydraulic station the oil creating a high pressure can move. As a result of a plate disperse and through inserts from thick plywood create powerful force.

Considered shells of revolution are composed of coaxial circular plates (they can be weakly warped) connected to the toroidal shell on the contour. In addition to the axial symmetry, the shell is symmetric with respect to the equidistant plates and, thus, only half of the construction can be considered (e.g. upper part). A prototype for such a construction can be found in [62]. The prototype upper part has meridian composed of three sections: segment of straight line and concave and convex arcs of the circles of the different radii.

The construction and working conditions of flat lifting jacks generate a number of problems of the mathematical model operation interesting to examinations [51]. One of directions is geometry optimisation. On paper, alternatives of the rotationally symmetric shells modeling FJ are analysed. Tensions of envelopes of different shapes are compared at the initial loading stage.

Denote main points on the radial axis, r_0 is the radius of the cylindrical pin, r_1 is the radius of the plate, r_2 is the point of the arcs conjugation, r_m is the point of the toroid's maximal ordinate, and r_3 is the external radius of the shell middle surface. Then, the prototype has the following sections of the meridian: segment of straight line $r \in [r_0, r_1]$, concave arc of the circle of the radius R_1 on the $r \in [r_1, r_2]$, convex arc of the circle of the radius R_2 on the $r \in [r_2, r_3]$. The main drawback of this form is a curvature jump followed by change of sign, leading to the stress concentration in the neighborhood of the arcs joint during loading process.

As an alternative, let us consider versions of the shells in which the meridian is composed of segments of the well-known functions [63]:

(1) segment of straight line on the $r \in [r_0, r_1]$; Cassini oval on the $r \in [r_0, r_1]$;
(2) exponent on the $r \in [r_0, r_m]$, concave arc of the circle of the radius R_2 on the $r \in [r_m, r_3]$;
(3) witch of Agnesi on the $r \in [r_0, r_m]$, concave arc of the circle of the radius R_2 on the $r \in [r_m, r_3]$;
(4) witch of Agnesi on the $r \in [r_0, r_m]$, limacon of Pascal on the $r \in [r_m, r_3]$;
(5) versions of the single curved given by power, fractional, exponential, or logarithmic functions, $r \in [r_0, r_3]$.

It should be noted that equations for curves are used not in classical form but with modified coefficients controlling the shape and scales through coordinates. Selection of these parameters is made in such a way that overall dimensions remain the same, i.e. diameter is $D = 2(r_3 + 1.5) = 520\,mm$ and toroidal section height is $H_T = 39\,mm$. Other measures have the magnitudes

$$r_0 = 8, \quad r_1 = 204.3, \quad r_m = 239, \quad r_3 = 258.5,$$

$$R_1 = 24.4, \quad R_2 = 19.5, \quad h_1 = 3,$$

where h_1 is shell thickness. In the future, it changes to the dimensionless quantities, where $R_* = r_1$, $h_* = h_1$.

The purpose of the analysis is to find the curves for which the following improvements (optimization criteria) will be made: (a) the number of sections is small; (b) behavior of the meridian curvature is as smooth as possible; and (c) the stresses in the toroidal part of the shell are reduced.

Particular emphasis is placed on the toroidal part of the shell because of its deformation duty complexity compared with the central part, which serves as platform attached to the rigid insert and deforms in its plane as a membrane during the main stage of the lifting. The toroidal part experiences tangential as well as bending stresses.

Consider version (1) based on the equation

$$(k_x x^2 + k_y y^2)^2 - 2c_k^2 (g_x x^2 - g_y y^2) - A_C = 0, \qquad (7.2.1)$$

where $A_C = a_k^4 - c_k^4$. The left part represents odd function with respect to x and y, i.e. it is symmetric with respect to coordinate axis. The central point $(x = 0)$ has horizontal tangent to the curve (7.2.1), and it gives an opportunity to make smooth conjugation between the toroid and the plate.

Then we will obtain the equation of the curve in the implicit form, given with respect to the vertical ordinate. By assuming $Y = y^2$ and introducing denotations

$$A_{Kas}(x) = (c_k^2 g_y + k_x k_y x^2)/k_y^2,$$
$$C_{Kas}(x) = (A_C + 2c_k^2 g_x x^2 - k_x^2 x^4)/k_y^2,$$

on the basis of (7.2.1), one can obtain equation

$$Y^2 + 2A_{Kas}(x)Y - C_{Kas}(x) = 0.$$

The necessary solution is given by

$$Y_{K1}(x) = -A_{Kas}(x) + \sqrt{[A_{Kas}(x)]^2 + C_{Kas}(x)}.$$

Let us transfer to the cylindrical coordinates (r, Z) and make scaling and shear of the curve along axis r. Then we have the function

$$Z_{0K}(r) = K_{Mz} \cdot \sqrt{Y_{K1}\left(\frac{r - r_1}{K_{Mr}}\right)}. \qquad (7.2.2)$$

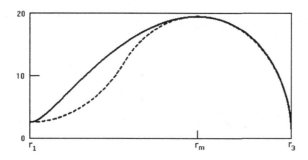

Fig. 7.2.1. View of curve (7.2.2) — solid line; dashed line — prototype.

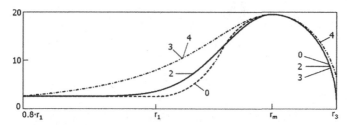

Fig. 7.2.2. Graphs of curves (2)–(4); the numbers correspond to the option number.

The condition of the overall equivalence is satisfied using the following set of the parameters:

$$K_{Mz} = 15.9, \quad K_{Mr} = 18.8, \quad c_k = 1.423, \quad a_k = 1.445,$$
$$k_x = 1.0, \quad k_y = 0.5, \quad g_x = 2.02, \quad g_y = 1.8.$$

The curve (7.2.2) is shown in Fig. 7.2.1 by the solid line, while the dashed line represents prototype.

A similar approach is used to build other curves. For the versions (2)–(4), the corresponding diagrams can be found in Fig. 7.2.2, where the figures denote the version number, while "0" corresponds to the prototype.

In versions (5), few curves were analyzed, two of which are given by

$$f_1(r) = A_{1z} \cdot \left[\lambda_{10} - \lambda_{11} \left(\frac{r - r_{1c}}{A_{1r}} \right)^2 \right] \cdot \exp[-\lambda_{1e}(r_{1c} - r)]$$
$$\times \log \left(\frac{r_{t1c} - r}{A_{1r}} \right) + B_{1z}, \tag{7.2.3}$$

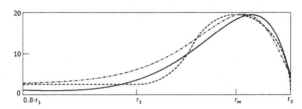

Fig. 7.2.3. Meridian graphs: solid line — formula (7.2.3); dash-dotted line — formula (7.2.4); dashed line — prototype.

$$A_{2z} = 0.48, \quad \lambda_{20} = 2.0, \quad \lambda_{21} = 0.12, \quad A_{2r} = 0.8,$$
$$r_{1c} = 230, \quad B_{1z} = 3, \quad \lambda_{1e} = 0.05.$$

$$f_2(r) = \frac{-A_{2z}}{r - r_{2c}} \cdot \left(\frac{r_{2c} - r}{A_{2r}} - \lambda_{20} \right)^{n_c} \exp[-\lambda_{21}(r_{2c} - r)^{n_e}]$$

$$\times \log \left(\frac{r_{2c} - r}{A_{2r}} - \lambda_{20} \right) + B_{2z} \qquad (7.2.4)$$

$$A_{2z} = 0.48, \quad \lambda_{20} = 2.0, \quad \lambda_{21} = 0.12, \quad A_{2r} = 0.8,$$
$$r_{2c} = 264, \quad B_{2z} = 2.5, \quad n_e = 0.95, \quad n_c = 2.65.$$

These curves are given in Fig. 7.2.3: solid and dash-dotted lines correspond to (7.2.3) and (7.2.4), respectively, while dashed line represents the prototype again.

Similar curves can be obtained by resetting of the parameters in (7.2.3) and (7.2.4) and ignoring logarithmic functions.

If the meridian ordinate is give by $Z(r)$, them Lame's coefficient A_1, slope angle of the normal to the revolution axis Φ_0, principle curvature k_1, k_2, and function ψ are determined by the following expressions:

$$A_1(r) = \sqrt{1 + (dZ(r)/dr)^2}, \quad \Phi_0(r) = -\mathrm{arctg}(dZ(r)/dr),$$
$$k_1(r) = [d\Phi_0(r)/dr]/A_1(r), \quad k_2(r) = \sin[\Phi_0(r)]/r,$$
$$\psi(r) = \cos[\Phi_0(r)]/r.$$

Some of the considered curves on the edge can have tangents close to vertical. Therefore, it is reasonable to use arc length s as an independent coordinate instead of polar radius; so, equations of the meridian and all shells parameters should be functions of s. This can be easily done analytically for the initial shell composed of circle arcs. For curves that are more complicated, a numerical approach should be used.

Let us consider an integral with variable upper limit, which denotes arc length as a function of polar radius

$$s(r) = \int_{r_0}^{r} \sqrt{1 + Z_{,r}(r)} \, dr. \qquad (7.2.5)$$

Let us evaluate s_i on the grid of values of $r_i, i = 0, \ldots, N$ using (7.2.5). Then, by using s_i as independent variables, the function $r(s)$ can be obtained by means of locally linear or spline interpolation. In replacing r by $r(s)$, we will end up with an equation for the meridian as function of s: $z(s) = Z(r(s))$ (replacement of r by $r(s)$ is also made in the functions of the geometry (7.2.5)).

The material of the shell is assumed to be homogeneous (steel) with modules E and ν. In transferring the problem to the dimensionless form, the following characteristic parameters were used: $E_* = E$, $\nu_* = \nu$, $R_* = r_1$, $h_* = h_1$.

7.3. Analysis of Elastic Stress State

In the problems of the radial axisymmetric deformation of the shells, all functions depend on the meridian coordinate α_1 only. To solve such boundary value problems, equations of the axisymmetric stress–strain state of the thin elastic revolution shells in the framework of the geometrically nonlinear theory of the quadratic approximation is used [61]. In the general shell theory, trihedral of the middle surface is considered as a special reference frame. Displacements and strains are related in this coordinate system by

$$\varepsilon_{11}(\alpha_1, z) = E_{11}(\alpha_1) + zK_{11}(\alpha_1),$$

$$\varepsilon_{22}(\alpha_1, z) = E_{22}(\alpha_1) + zK_{22}(\alpha_1),$$

$$E_{11} = u' + k_1 w + \vartheta_1^2/2; \quad E_{22} = \psi u + k_2 w,$$

$$\vartheta_1 = -w' + k_1 u, \quad K_{11} = -\vartheta_1', \quad K_{22} = \psi \vartheta_1,$$

$$\psi = A_2'/A_2, (\ldots)' = d(\ldots)/(A_1 d\alpha_1). \qquad (7.3.1)$$

Here, u, v are middle surface displacements; ϑ_1 is rotation angle of the normal; A_1, A_2 are Lame coefficients; and k_1, k_2 are the main curvature. Positive direction of the normal is assumed to be in outward direction with respect to the shell.

Equilibrium equations for the forces and moments result from the Lagrange principle based on the kinematic relations (7.3.1)

$$T_{11}' + \psi(T_{11} - T_{22}) + k_1 Q_{11} + q_1 = 0,$$
$$Q_{11}' + \psi Q_{11} - k_1 T_{11} - k_2 T_{22} + q_3 = 0,$$
$$M_{11}' + \psi(M_{11} - M_{22}) - Q_{11} - T_{11}\vartheta_1 = 0. \tag{7.3.2}$$

Constitutive equations for the isotropic shells are given by

$$T_{11} = B(E_{11} + \nu E_{22}), \quad T_{22} = B(E_{22} + \nu E_{11}),$$
$$M_{11} = D(K_{11} + \nu K_{22}), \quad M_{22} = D(K_{22} + \nu K_{11}),$$
$$B = Eh/(1 - \nu^2), \quad D = Eh^3/[12(1 - \nu^2)], \tag{7.3.3}$$

B, D, G_{13} are effective shell stiffness for tension-compression, bending and transverse shear, respectively; Young's modulus, and ν is Poisson's ratio.

Stresses and stress intensity can be expressed as

$$\sigma_{11} = E_\nu(\varepsilon_{11} + \nu\varepsilon_{22}), \quad \sigma_{22} = E_\nu(\varepsilon_{22} + \nu\varepsilon_{11}),$$
$$E_\nu = E/(1 - \nu^2), \quad \sigma = (\sigma_{11}^2 + \sigma_{22}^2 - \sigma_{11}\sigma_{22})^{1/2}. \tag{7.3.4}$$

Boundary conditions are found from the requirement stating that contour integrals of the Lagrange functional are equal to zero. In the case of the homogeneous boundary conditions, generalized displacements or generalized strains are equated to null

$$u(1 - \gamma_1) + \gamma_1 T_{11} = 0, \quad w(1 - \gamma_2) + \gamma_2 Q_{11} = 0,$$
$$\vartheta_1(1 - \gamma_3) + \gamma_3 M_{11} = 0, \quad \alpha_1 = \alpha_{11};$$
$$u(1 - \gamma_4) + \gamma_4 T_{11} = 0, \quad w(1 - \gamma_5) + \gamma_5 Q_{11} = 0,$$
$$\vartheta_1(1 - \gamma_6) + \gamma_6 M_{11} = 0, \quad \alpha_1 = \alpha_{12}. \tag{7.3.5}$$

Here, coefficients γ_j are assumed to be 0 or 1 and allow for the setting of different boundary conditions.

Let us introduce main functions

$$y_1 = T_{11}, \ y_2 = Q_{11}, \ y_3 = M_{11}, \ y_4 = u, \ y_5 = w, \ y_6 = \vartheta_1 \quad (7.3.6)$$

and establish a canonical system of the differential equations of the sixth-order for them

$$\mathbf{y}' = \mathbf{f}(\alpha_1, \mathbf{y}) + \mathbf{p}_1(\alpha_1) = 0$$
$$\mathbf{y} = (y_1, \ldots, y_6), \quad \mathbf{f} = (f_1, \ldots, f_6),$$
$$\mathbf{p} = (-q_1, -q_3, 0, 0, 0, 0). \quad (7.3.7)$$

The right-hand sides of equations f_j are given by

$$f_1 = \psi(T_{22} - y_1) - k_1 y_2,$$
$$f_2 = -\psi y_2 + k_1 y_1 + k_2 T_{22},$$
$$f_3 = \psi(M_{22} - y_3) + y_2 + y_1 y_6,$$
$$f_4 = E_{11} - y_1 - k_1 y_5 - 0.5(y_6)^2,$$
$$f_5 = k_1 y_4 - y_6,$$
$$f_6 = K_{11}, \quad (7.3.8)$$

where

$$E_{22} = \psi u + k_2 w, \quad K_{22} = \psi y_6,$$
$$K_{11} = y_3/D - \nu K_{22}, \quad E_{11} = y_1/B - \nu E_{22},$$
$$T_{22} = B(E_{22} + \nu E_{11}), \quad M_{22} = D(K_{22} + \nu K_{11}). \quad (7.3.9)$$

In developing an algorithm, these equations are transformed into the dimensionless form using

$$\{U, W, u, w, h, z\}_U = \{\cdots\}_D/h_*,$$
$$\{A_1, A_2\}_U = \{\cdots\}_D/R_*,$$
$$\{k_1, k_2, K_{11}, K_{22}\}_U = R_*\{\cdots\}_D,$$
$$\{\vartheta_1, \varepsilon_{11}, \varepsilon_{22}, E_{11}, E_{22}\}_U = \{\cdots\}_D/\varepsilon_*,$$
$$\{\sigma_{11}, \sigma_{22}, \sigma, \sigma_{as}, \sigma_{ys}\}_U = \{\cdots\}_D/(E\varepsilon_*),$$

$$\{T_{11}, T_{22}, Q_{11}\}_U = \{\cdots\}_D R_*/A_*,$$

$$\{M_{11}, M_{22}\}_U = \{\cdots\}_D R_*/D_*,$$

$$\{q_1, q_3\}_U = \{\cdots\}_D R_*^2/A_*, \quad B_U = B_D/B_*,$$

$$D_U = D_D/D_*, \quad E_U = E_D/E_*, \qquad (7.3.10)$$

where

$$\varepsilon_* = h_*/R_*, \quad B_* = E_* h_*/(1 - \nu_*^2), \quad A_* = B_* h_*, \quad D_* = B_* h_*^2. \quad (7.3.11)$$

Normalization parameters E_*, ν_*, R_*, h_* have sense and dimension of the Young's modulus, Poisson's ratio, curvature radius, or linear measure and thickness, respectively. Dimensionless values are combined by braces with index "U" in the left parts of equations. Their dimension analogs are given in the braces with index "D" in the right parts of (7.3.10). Governing equations in the dimensionless form are given by

$$dy_j/d\alpha_1 = A_1 F_j(\alpha_1, \mathbf{y}), \quad j = 1, \ldots, 6; \qquad (7.3.12)$$

$$F_1 = \psi(T_{22} - y_1) - k_1 y_2 - q_1,$$

$$F_2 = -\psi y_2 + k_1 y_1 + k_2 T_{22} - q_3,$$

$$F_3 = \psi(M_{22} - y_3) + y_2/\varepsilon_* + y_1 y_6,$$

$$F_4 = E_{11} - y_1 - k_1 y_5 - 0.5\varepsilon_* y_6^2,$$

$$F_5 = k_1 y_4 - y_6, \quad F_6 = K_{11}/\varepsilon_*; \qquad (7.3.13)$$

$$E_{22} = \psi y_4 + k_2 y_5, \quad K_{22} = \varepsilon_* \psi y_6,$$

$$K_{11} = y_3/D - \nu K_{22}, \quad E_{11} = y_1/B - \nu E_{22},$$

$$T_{22} = B(E_{22} + \nu E_{11}), \quad M_{22} = D(K_{22} + \nu K_{11}). \quad (7.3.14)$$

For simplicity sake, index "U" is not used in the following expressions. In the closed form, the right parts of (7.3.13) of the system (7.3.12) are fully determined through independent functions y_j, and for an isotropic shell, it is expressed as

$$F_1 = \psi[(\nu - 1)y_1 + B(1 - \nu^2)(\psi y_4 + k_2 y_5)] - k_1 y_2 - q_1,$$

$$F_2 = -\psi y_2 + (k_1 + \nu k_2)y_1 + k_2 B(1 - \nu^2)(\psi y_4 + k_2 y_5) - q_3,$$

$$F_3 = \psi[(\nu - 1)y_3 + D(1 - \nu^2)\varepsilon_* \psi y_6] + y_2/\varepsilon_* + y_1 y_6,$$

$$F_4 = y_1/B - \nu\psi y_4 - (k_1 + \nu k_2)y_5 - 0.5\varepsilon_*(y_6)^2,$$

$$F_5 = k_1 y_4 - y_6, F_6 = y_3/(\varepsilon_* D) - \nu\psi y_6. \qquad (7.3.15)$$

Boundary value problems for the system of equation (7.3.12) with right-hand sides of equations (7.3.15) were considered with the following boundary conditions: left edge (in the central part of the shell) is allowed to move vertically; right edge is clamped. Boundary value problem was solved by shooting method.

Two types of the loading by internal pressure have been considered — $q_1 = 0$, $q_3 \neq 0$. Type A — pressure of intensity $0.1\,MPa$ (dimensionless value is $q_3 = 0.002$) is applied to the whole surface of the shell; type B — pressure of intensity $0.6\,MPa$ (dimensionless value is $q_3 = 0.012$) is applied to the toroidal part, only $r_1 < r < r_3$. These types simulate initial stage of the choosing of the clearance between weight and shell, and in some way stage of the intimate contact of the shell with insert taking into account resistance of the construction, which, in turn, compensates internal pressure in the platform zone. Thus, for the type of loading B central part of the shell works almost without bending.

To illustrate this, Fig. 7.3.1 provides information on the stress intensity σ (7.3.4) for the toroidal part of the shells of version 1 with Cassini oval (solid curves) compared with a prototype shell

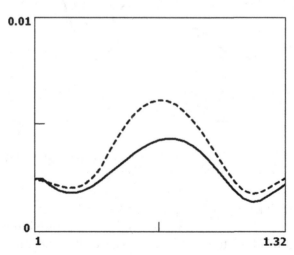

Fig. 7.3.1. Stress intensity σ on the middle surface under type B loading: solid curve — Cassini oval; dashed line — prototype.

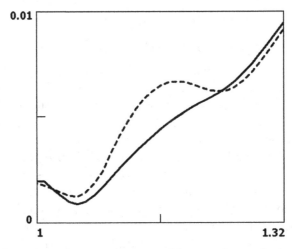

Fig. 7.3.2.   Stress intensity σ on the external front surface under type B loading: solid curve — Cassini oval; dashed line — prototype.

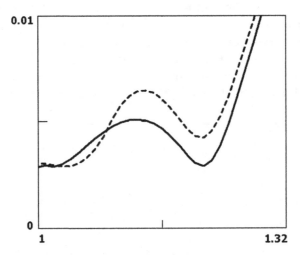

Fig. 7.3.3.   Stress intensity σ on the inner front surface under type B loading: solid curve — Cassini oval; dashed line — prototype.

(dashed line). The type of loading is B. Figure 7.3.1 shows stress intensity on the middle surface. Figure 7.3.2 shows stress intensity on the the external front surface, and Fig. 7.3.3 on the inner front surface.

Shells of version (1) and the prototype have similar locations of dangerous points, but the stresses for type (1) are significantly reduced. The same properties are demonstrated for version (2). That's why analysis of different forms is of practical relevance for the safety factor. It is clear that the considered shell for large displacements and strains works out of the scope of elasticity theory. It would be interesting to take into account these factors in a more complicated model in the further research. However, experiments under prototype shell show that the most stressed zones discovered in the scope of linear elasticity remain the most stressed zones when the loading is increased. During cyclic work of lifting, construction is forced to back to the nominal initial state, and this repetition exposes the material to storage low-cycle fatigue (microcracks) causing gradual implantation damages in the construction. Hence, relatively simple solutions of the elastic problem have proven to be useful in providing information on possible zone of the damage.

7.4. Simulation of Flat Jack Operation in Case of Large Displacements

The shells of rotation formed by two coaxial and parallel round plates conjugated on the contours with a torus cut along the inner diameter are considered. The design is symmetrical with respect to the horizontal plane and equidistant with respect to the plates. The Meridian of the torus has a composite geometry and is formed by arcs of smooth curves, in particular, circles. At the points of conjugation of elements, there are discontinuities of curvature with change of sign. When loading the shell with internal hydraulic oil pressure, large displacements (a few centimeters) and significant thrust forces (tens of tons) are provided with resistance on the plates. Shells of this type are modeled structures called flat jacks. Such power mechanisms are used in the NPO "Interbiotech" for lifting and leveling of high-rise buildings that have received a roll in operation during the subsidence of the soil. The relevance of the analysis of such structures is due to the lack of methods for their calculation, the need to assess the options for jacks with different geometric and technical characteristics, as well as due to the advantages of flat jacks compared to pistons. The report deals with a geometrically and physically nonlinear mathematical model based on the equations of shell theory for large displacements, angles of rotation and deformations, and the relations determining the properties of the material, in which the plasticity diagram for logarithmic deformations is used. In the applied semi-analytical semi-inverse method, the geometry and thickness are set so that there is selection of parameters with sufficient accuracy to satisfy the set of equations of shell theory. To do this, we use the formula derived from the equations that relate the hydrostatic pressure to the parameters of the problem. One of the schemes of construction and calculation algorithm is implemented when the plates are rigid and the torus meridian changes geometry without elongation. Tension in the circumferential direction and thickness changes are taken into account.

The model of the shell type of flat jack in deformations close to the limit is considered [55]. To solve this problem, we also apply the semi-inverse method, when, *a priori*, the shape of the meridian of the obtained shell and the reaction of bonds are given. The application of this method is similar to the modification in Section 6.7. Given the flexibility of the shell along the meridian, it is assumed that its stretching is small, that is, $\varepsilon_1 \ll 1$. The calculation scheme of modeling of flat jacks used for lifting and leveling of heavy structures [50] is considered.

Flat jack can be considered as a kind of mechanism for generating large forces and movements. The device is formed by two coaxial parallel round plates, which are conjugated on the contours with a torus cut along the inner diameter. The design is also symmetrical with respect to the horizontal plane and equidistant with respect to the plates. The meridian of the torus has a composite geometry and is formed by arcs of smooth curves, in particular, circles. At the points of contact of the elements of the torus there are breaks of curvature with a change of sign.

Here, the task is a set of mathematical modeling of the work of a flat jack in the range of large displacements, up to the maximum possible. This requires the use of geometrically and physically nonlinear equations. It is necessary to take into account the changes in the metric of the shell due to the appearance of large deformations and inelastic behavior of the material.

The applied algorithm reduces the problem to the conclusion of some functions to a constant depending on the vector of variable parameters. The constant is equal to the value of uniform pressure, under the action of which the shell is deformed to a given level of characteristic displacement. With a small number of parameters, they can be selected interactively.

The chapter uses a geometrically and physically nonlinear mathematical model based on the equations of shell theory for large displacements, angles of rotation and deformations, and defining the properties of the material relations, in which the plasticity diagram for logarithmic deformations is used.

We present a summary of the basic equations of the model. Kinematic relations are as follows:

$$e_1 = (\varepsilon_1 + \zeta \kappa_1), \quad e_2 = (\varepsilon_2 + \zeta \kappa_2)$$

$$\varepsilon_1 = (w' \sin \Phi + u' \cos \Phi)/\alpha_o + \cos(\Phi - \Phi_o) - 1, \quad \varepsilon_2 = u/r_o$$

$$\kappa_1 = \Phi'_o/\alpha_o - (1 + \varepsilon_3)K_1, \quad \kappa_2 = (\sin \Phi_o)/r_o - (1 + \varepsilon_3)K_2, \quad (7.4.1)$$

$$K_1 = \Phi'/\alpha_o, \quad K_2 = (\sin \Phi)/r_o; \quad (7.4.2)$$

$$\gamma_1 = \gamma_o/(1 + \varepsilon_1), \quad \gamma_o = (w' \cos \Phi - u' \sin \Phi)/\alpha_o - \sin(\Phi - \Phi_o).$$

$$(7.4.3)$$

All functions depend on the curvilinear coordinate $\alpha_1 = \xi$. Here, the prime denotes the derivative with respect α_1; α_o is Lame coefficient of the initial shell; Φ_o is the angle between the normal to the reference shell and the z-axis; Φ is the angle between the normal to the deformed membrane and the z-axis; $\delta_j = 1 + \varepsilon_j, \, j = 1, 2, 3$. The incompressibility condition is assumed to be satisfied: $\delta_1 \delta_2 \delta_3 = 1$, $\delta_3 = \delta_1^{-1} \delta_2^{-1}$.

Equilibrium equations are

$$(r_o \bar{V}^o)' + \alpha_o r_o p_w^o = 0; \quad (7.4.4)$$

$$(r_o \bar{H}^o)' - \alpha_o \bar{N}_2^o + \alpha_o r_o p_u^o = 0; \quad (7.4.5)$$

$$(r_o M_1^o)' - \alpha_o M_2^o \cos \Phi - \alpha_o r_o \delta_1 Q^o = 0. \quad (7.4.6)$$

$$p_w^o = \delta_1 \delta_2 p_w, \quad p_u^o = \delta_1 \delta_2 p_u. \quad (7.4.7)$$

The internal forces $\bar{V}^o$ and $\bar{H}^o$ are oriented vertically (along the axis of symmetry) and horizontally (along the radius of the cylindrical coordinate system), respectively, and Q^o is the shearing force

$$\bar{V}^o = \bar{N}_1^o \sin \Phi + Q^o \cos \Phi,$$

$$\bar{H}^o = \bar{N}_1^o \cos \Phi - Q^o \sin \Phi. \quad (7.4.8)$$

Values $\bar{N}_1^o, \bar{N}_2^o$ are generalized forces associated with tangential internal forces N_1^o, N_2^o and moments M_1^o, M_2^o

$$\bar{N}_1^o = N_1^o + \delta_1^{-1}(K_1 M_1^o + K_2 M_2^o),$$
$$\bar{N}_2^o = N_2^o + \delta_2^{-1}(K_1 M_1^o + K_2 M_2^o),$$
$$N_1^o = \delta_2 N_1, \quad N_2^o = \delta_1 N_2, \quad Q^o = \delta_2 Q,$$
$$M_1^o = \delta_2 M_1, \quad M_2^o = \delta_1 M_2. \tag{7.4.9}$$

Internal pressure is considered as a load. Given that the positive direction of the normal inside the shell, we have

$$p_w = -p\cos\Phi, \quad p_u = p\sin\Phi. \tag{7.4.10}$$

The defining relations of the Davis–Nadai type are accepted in the form

$$N_1^o = \bar{B}_1(\bar{\varepsilon}_1 + 0.5\bar{\varepsilon}_2),$$
$$N_2^o = \bar{B}_2(\bar{\varepsilon}_2 + 0.5\bar{\varepsilon}_1),$$
$$M_1^o = \bar{D}_1\left(\bar{\kappa}_1 + 0.5\bar{\kappa}_2\right),$$
$$M_2^o = \bar{D}_2\left(\bar{\kappa}_2 + 0.5\bar{\kappa}_1\right); \tag{7.4.11}$$
$$\bar{\varepsilon}_1 = \ln(\delta_1), \quad \bar{\varepsilon}_2 = \ln(\delta_2),$$
$$\bar{\kappa}_1 = \kappa_1/\delta_1, \quad \bar{\kappa}_2 = \kappa_2/\delta_2; \tag{7.4.12}$$
$$\bar{B}_1 = \bar{B}/\delta_1, \quad \bar{B}_2 = \bar{B}/\delta_2,$$
$$\bar{D}_1 = \delta_3^2\delta_2\bar{D}, \quad \bar{D}_2 = \delta_3^2\delta_1\bar{D},$$
$$\bar{B} = (4/3)\Lambda(\bar{\varepsilon})h_o, \quad \bar{D} = (1/9)\Lambda(\bar{\varepsilon})h_o^3. \tag{7.4.13}$$

Here $\Lambda(\bar{\varepsilon}) = C\bar{\varepsilon}^{\eta-1}$ is the secant modulus of material plasticity diagram, where

$$\bar{\varepsilon} = (2/\sqrt{3})\sqrt{\bar{\varepsilon}_1^2 + \bar{\varepsilon}_1\bar{\varepsilon}_2 + \bar{\varepsilon}_2^2}$$

is the intensity of the logarithmic strain.

The material diagram can be identified by characteristic values, which are usually given in reference books. These are the modulus of elasticity E, the elastic limit σ_{02}, the limit elastic deformation $\varepsilon_{02} = \sigma_{02}/E$, the limit stress intensity σ_l, and the limit intensity of logarithmic deformations $\bar{\varepsilon}_l$. These data define the parameters of the exponential approximation of the diagram with nonlinear material properties

$$\sigma(\bar{\varepsilon}) = C\bar{\varepsilon}^\eta,$$

where

$$\eta = [\ln(\sigma_l) - \ln(\sigma_{02})]/[\ln(\bar{\varepsilon}_l) - \ln(\varepsilon_{02})], \quad C = \sigma_l/\bar{\varepsilon}_l^\eta.$$

The secant module was set to E when $\bar{\varepsilon} < \varepsilon_{02}$ and $\Lambda(\bar{\varepsilon})$ when $\bar{\varepsilon} \geq \varepsilon_{02}$.

We integrate equations (7.4.4) and (7.4.5). Given (7.4.8), we obtain

$$r_0(\bar{N}_1^o \sin \Phi + Q^o \cos \Phi) = p \cdot I_{11} + C_V, \qquad (7.4.14)$$

$$r_0(\bar{N}_1^o \cos \Phi - Q^o \sin \Phi) = I_{21} - p \cdot I_{22} + C_H, \qquad (7.4.15)$$

where C_V and C_H are integration constants associated with vertical and horizontal forces on the contour $\xi = \xi_0$; I_{11}, I_{21}, I_{22} are integrals with variable upper limits

$$I_{11} = \int_{\xi_0}^{\xi} \alpha_o r_o \delta_1 \delta_2 \cos \Phi d\xi, \quad I_{21} = \int_{\xi_0}^{\xi} \alpha_o \bar{N}_2^o r_o d\xi,$$

$$I_{22} = \int_{\xi_0}^{\xi} \alpha_o r_o \delta_1 \delta_2 \sin \Phi d\xi. \qquad (7.4.16)$$

From the relations (7.4.14) and (7.4.15), we express Q^o. To do this, we perform the operation

$$r_o^{-1}[(7.4.14) \cdot \cos \Phi - (7.4.15) \cdot \sin \Phi].$$

We get

$$Q^o = r_o^{-1}[(p \cdot I_{11} + C_V) \cos \Phi$$

$$- (I_{21} - p \cdot I_{22} + C_H) \sin \Phi]. \qquad (7.4.17)$$

We substitute the expression (7.4.17) for Q^o in equation (7.4.6) and integrate it. We get

$$r_o M_1^o = I_{31} + p \cdot (J_{11} + J_{22}) + C_V J_{31}$$

$$- J_{32} - C_H J_{33} + C_M, \qquad (7.4.18)$$

where

$$J_{11} = -\int_{\xi}^{r_b} \alpha_o \delta_1 I_{11} \cos \Phi d\xi, \quad J_{22} = -\int_{\xi}^{r_b} \alpha_o \delta_1 I_{22} \sin \Phi d\xi,$$

$$J_{31} = -\int_{\xi}^{r_b} \alpha_o \delta_1 \cos \Phi d\xi, \quad J_{32} = -\int_{\xi}^{r_b} \alpha_o \delta_1 I_{21} \sin \Phi d\xi,$$

$$J_{33} = -\int_{\xi}^{r_b} \alpha_o \delta_1 \sin \Phi d\xi. \tag{7.4.19}$$

C_M is the integration constant associated with the moment reaction on the contour.

The pressure p does not depend on ξ and is taken out beyond the integral sign. From (7.4.18), it follows that

$$p = \frac{r_0 M_1^o + J_{32} - J_{31} - C_M - C_V J_{31} + C_H J_{33}}{J_{11} + J_{22}}. \tag{7.4.20}$$

Numerators and denominators of the functional (7.4.20) depend on ξ. A further problem is to deduce the pressure p by a value close to the constant value with acceptable accuracy. The control levers here are the parameters of the approximation of the shell shape and the integration constant. This can be done interactively in a numerical experiment. In this case, of course, the physical relations (7.4.11)–(7.4.13) as well as the condition of incompressibility of the material are involved.

We introduce the notation of the main parameters of the shell in the initial state, Fig. 7.4.1: $h_p = h_1$ is the thickness of the plate;

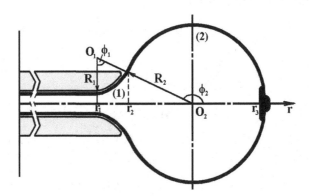

Fig. 7.4.1.   Construction scheme.

r_3 is the outer radius of the middle surface of the torus; 2δ and $2\delta_1 = 2\delta + h_1$ is the distance between the plates and their middle surfaces; $r_p = r_1$ radius; R_2 is the radius of the arc 2 forming the convex part of the torus (section 2); $r_{O1} = r_1$ and $r_{O2} = r_3 - R_2$ are the coordinates of the centers of circles, arcs of which form the toroidal part; and ϕ_1 and ϕ_2 are coverage angles of arcs of circles

$$r_2 = r_{O2} - R_2 \sin(\phi_2 - \pi/2), \quad z_2 = R_2 \cos(\phi_2 - \pi/2)$$

are the coordinates of the point of junction between the arcs. The radius of the arc 1 forming the concave part of the torus (section 1) and the mating plate with part 2 of the torus is calculated by the formula

$$R_1 = 0.5.[(z_2 - \delta_1)^2 + (r_2 - r_1)^2]/(z_2 - \delta_1).$$

The thickness of the shell of the torus changes from h_1 at the point $r = r_p$ to $k_h h_1$, $k_h < 1$, at the point $r = r_3$.

The equations of the sections of the middle surface of the composite shell will be as follows: plate $Z_p(r) = \delta_1$; the torus in sections 1 and 2

$$Z_1(r) = R_1 + \delta_1 - [(R_1)^2 - (r - r_1)^2]^{1/2}, \quad Z_2(r) = [(R_2)^2 - (r - r_{O2})^2]^{1/2}.$$

A single surface $Z_o(r)$ is formed by a superposition of piecewise defined functions defined by the corresponding equations on its interval and vanishing outside it.

When using the shell for its intended purpose, it is placed in a package between the inserts, which are round plates of thick plywood. With a significant load of resistance of the raised structure, the inserts prevent buckling of the metal plate, leaving it flat in the normal lifting mode.

From the structural features of the shell, we note the presence of a sufficiently rigid ring on the outer contour, on which the upper and lower parts are connected by welding. Therefore, when considering one (upper) half, we can assume that the fixing of the shell is stiff.

We consider the plate separately under the action of the tensile load on its contour $r = r_1$. We use the homogeneity of the stress–strain state and the condition of incompressibility of the material.

During the operation of the system, plastic deformation appear, s but not so large that it was necessary to use the logarithmic relative elongations. Taking these factors into account, it is relatively easy to obtain a connection between the boundary load N_1 and the displacement u_1 arising on the plate contour.

The formula looks like this

$$N_{1p} = C(2u_1/r_1)^{\eta-1}[2u_1h_1/(r_1 + u_1)], \qquad (7.4.21)$$

where C and η are material constants. For material type steel 3, $C = 441.01286$, $\eta = 0.10741$. The plate works by stretching in its plane and effectively resists deformation. In the calculation scheme, we take into account the displacement of both the contour of the plate (u_1) and the external diameter of the structure, u_3.

We are interested in the height and weight of the load, close to the maximum possible jacks. The maximum height is determined by the movement of the inner contour of the toroidal part of the shell. We consider it separately, setting the power characteristics on the inner circuit. These values refer to the functional control parameters (7.4.20).

We assume that the point of joining the torus to the plate undergoes displacement w_p at the transition of the shell from the initial state (0) to the deformed (d), Fig. 7.4.2.

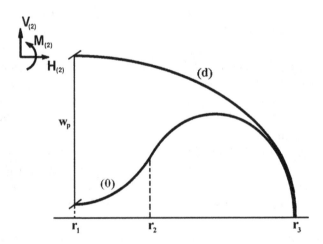

Fig. 7.4.2. Shell meridian before (0) and after (d) deformation.

To approximate the deformed surface, it is preferable to use a single equation that provides continuity of the Meridian curvature change. The use of an elliptical shape is acceptable.

$$Z_d(r_d) = (b_{d2}/a_{d2})\sqrt{a_{d2}^2 - (r_d - r_1)^2}. \qquad (7.4.22)$$

Since the change in the shape of the torus zone occurs by bending, it can be assumed that the elongation along the meridian is small: $\varepsilon_1 \approx 0$. Then the lengths of the meridian arcs of the shell are preserved. The arc length before deformation is $S_0 = S_1 + S_2$, where $S_1 = R_1\phi_1$, $S_2 = R_1\phi_2$.

The length of the arc of the meridian of the deformed shell is

$$S_d = \int_{r_1}^{r_3} \sqrt{1 + (d(Z_d(r_d))/dr_d)^2} dr_d. \qquad (7.4.23)$$

The parameters of the curve (22) are chosen by the condition $S_d \approx S_0$. Also, the coordinates of s and s_d points coincide when moving in an arc, which means that s_d is Lagrangian. Therefore, it is further advisable to move to the independent coordinate s.

The polar radius of the initial middle surface of the convex torus r_o as a function of the current point of the meridian arc is given as

$$r_{01}(s) = r_1 + R_1 \sin[(s - s_1)/R_1], \quad s \in [s_1, s_2], \qquad (7.4.24)$$

$$r_{02}(s) = r_2 + R_2(k_r - \cos[\pi/2 - \xi_{12} + (s - s_2)/R_2], \quad s \in [s_2, s_3], \qquad (7.4.25)$$

where $k_r = (r_3 - r_2)/R_2 - 1$, $s_1 = r_1$, $s_2 = s_1 + S_1$, $s_3 = s_2 + S_2$.

Functions (7.4.24) and (7.4.25), specified piecewise, can be combined into $r_0(s)$ using logical operators. When counting the arc, the Lamé coefficient is $\alpha_0 = 1$. The equation of the meridians, depending on the s, will be $Z_{0s}(s) = Z_0(r_0(s))$, $Z_{ds}(s) = Z_d(r_0(s))$. The angles of inclination of the normals are determined by the formulas

$$\Phi_0(s) = -\text{arctg}\left(\frac{d}{ds}Z_{0s}(s)\right), \quad \Phi(s) = -\text{arctg}\left(\frac{d}{ds}Z_{ds}(s)\right). \qquad (7.4.26)$$

From the relations (7.4.1) and (7.4.2), it is possible to obtain the equation

$$u' = \alpha_0[(1 + \varepsilon_1)\cos\Phi - \gamma_0 \sin\Phi - \cos\Phi_0)]. \qquad (7.4.27)$$

which is a condition of for compatibility of deformations. Considering that $\varepsilon_1 \approx 0$, $\gamma_0 \approx 0$, we have a simplified version

$$u' = \alpha_o(\cos \Phi - \cos \Phi_o). \qquad (7.4.28)$$

Since the initial and deformed forms of the shell are given, this ratio is integrated

$$u(s) = \int_{r_1}^{s} \alpha_o(s)[\cos(\Phi(s)) - \cos(\Phi_o(s))]ds + u_1. \qquad (7.4.29)$$

Then the relative elongation $\varepsilon_{22}(s) = u(s)/r_o(s)$ is determined. The incompressibility of the material $(1 + \varepsilon_{22}(s))(1 + \varepsilon_{33}(s)) = 1$ implies

$$\varepsilon_{33}(s) = (1 + \varepsilon_{22}(s))^{-1} - 1.$$

In this case, the intensity of deformation is $\varepsilon = (2/\sqrt{3})\varepsilon_{22}$.

Numerators and denominators in the formula (7.4.20) for p functions from s. Controlling the parameters of the approximation of the shell shape and the integration constants, it is necessary to deduce the pressure p by a value close to the constant value with acceptable accuracy. This can be done interactively in a numerical experiment.

We consider a structure having dimensions (here and further in mm): $h_1 = 3$, outer diameter $D = 520$, $r_3 = (D - h_1)/2 = 258.5$; distance between plates $2\delta = 2$, $\delta = 1$; $r_1 = 408.65/2 = 204.325$, $r_2 = 223.63$, $z_2 = 12.005$. Arc coverage angles of circles: $\phi_1 = 52$ degrees, $\phi_2 = 142$ degrees. The material of construction is steel, with characteristics: $E = 0.21 \cdot 10^6\,MPa$, $\sigma_{02} = 210\,MPa$, $\sigma_l = 380\,MPa$, $\varepsilon_{02} = 0.001$, $\bar{\varepsilon}_l = 0.25$, $C = 441\,MPa$, $\eta = 0.107$.

We consider the lifting height of the plate $w_p = 28.55$, which makes it possible to fulfill the condition of preserving the lengths of the arcs. Then the vertical half-axis of the ellipse is $b_d = w_p + \delta + h_1/2 = 31.05$, horizontal half-axis is $a_d = (r_3 - r_1) + (u_3 - u_1) = 59.175$.

When the constant value is reached, p is the pressure that corresponds to the specified blow-up level. In this case, the pressure was about 10 megapascals (MPa).

The load rating is not carried out by simple calculation. To do this, it is necessary to involve techniques based on experimental data. This is due to the specific operation of the flat jack with an insert.

With a change in the value of the jack's stroke, the radius of the torus and the area of contact of the jack with the support insert change. Therefore, the load capacity of the jack decreases with increasing stroke. According to the technical method, taking into account these factors, the load capacity for this internal pressure is estimated at 1422 kilonewtons (kN).

Chapter 8

Equations for Axisymmetric Shells Calculating within the Elasticity Limits

Shells model the supporting or enclosing thin-walled elements of technical structures and buildings. In geometry, they are two-dimensional models. However, depending on the specific geometric features, types of loads, physical and mechanical properties of the material, the presence of contact with a continuous medium, the effects of physical fields, etc., the tasks of studying the behavior of shells can be based on models with one to four independent variables, including time.

The choice of certain models and equations of shell theory must be adequate to the degree of complexity of the problems to be solved and the required accuracy. Problems can be static, dynamic, stationary or non-stationary; temperature, geometrically and (or) physically linear and (or) nonlinear; problems of strength; stability; endurance; destruction; natural and forced vibrations; problems of elasticity, viscoelasticity, viscoelasticity; related or unrelated; and with other combinations of key words — characteristics. The multiplicity of tasks reflects the diversity of the purpose and functional purpose of shell structures and their working conditions under the influence of external factors.

To reduce non-dimensional boundary value problems to two-point boundary value problems, you must remove the "excess" independent coordinates. Common ways to do this are series expansions

179

based on these coordinates, finite difference approximations, spline approximations, and so on. In the class of shells of rotation and closed shells, trigonometric Fourier series on the circumferential coordinate are widely used.

By subsequent transformations, the resulting system of ordinary differential equations can be reduced to a canonical form relative to the resolving system of functions. Depending on the model used, the system of resolving functions may include components of generalized displacements and their derivatives, stress (effort) functions and their derivatives, forces, and moments.

In some cases, especially for shells of rotation, a mixed form of the resolving vector is very convenient, the coordinates of which contain an equal number of generalized kinematic and force components that are included in the natural boundary conditions and make up static–geometric pairs. This makes it easy to satisfy homogeneous boundary conditions and use a unified matrix form for writing them. Some of these systems of equations used for solving problems of calculating axisymmetric stress–strain states are presented in this chapter.

8.1. Quadratic-Nonlinear Equations in Natural Trihedron of the Median Surface

Axisymmetric shells under radially symmetric deformation conditions are considered. Enter a grid of curved coordinates α_1, α_2 on the lines of the main curvature of the median surface. Shells are considered relatively thin, for which the product of the main curvatures by the thickness is significantly less than one ($k_j h \ll 1$, $j = 1, 2$). The main assumptions of the quadratic theory of elastic shells of Kirchhoff–Love type in the version of V. V. Novozhilov [64, 65] are also assumed to be fulfilled. These include

- orthogonality of the material normal to the median surface during deformations (i.e. transverse shear deformations $\varepsilon_{13} \sim 0$) and smallness of normal stresses in thickness compared to tangential ones ($\sigma_{33} \sim 0$);
- the displacement that can be comparable to the thickness of the shell;

- the components of the strain tensor and the squares of the rotation angles that may be of the same order, but are small compared to one.

Since radial axisymmetric deformation is assumed, all functions depend only on α_1. In general shell theory, the mid-surface trihedron is usually used as a special reference system. Kinematic relations for meridional and normal displacements and components of tangential deformations in this system have the form [22, 65, 66]

$$U(\alpha_1, z) = u(\alpha_1) + z\vartheta_1(\alpha_1), \quad W(\alpha_1, z) = w(\alpha_1);$$

$$\varepsilon_{11}(\alpha_1, z) = E_{11}(\alpha_1) + zK_{11}(\alpha_1),$$

$$\varepsilon_{22}(\alpha_1, z) = E_{22}(\alpha_1) + zK_{22}(\alpha_1); \tag{8.1.1}$$

$$E_{11} = u' + k_1 w + \vartheta_1^2/2, \quad E_{22} = \psi u + k_2 w,$$

$$\vartheta_1 = -w' + k_1 u, \quad K_{11} = -\vartheta_1', \quad K_{22} = \psi \vartheta_1; \tag{8.1.2}$$

$$(\ldots)' = (\ldots)_{,\alpha_1}/A_1 = d(\ldots)/(A_1 d\alpha_1), \quad \psi = A_2'/A_2,$$

where

α_1, α_2 are curved orthogonal coordinates of the reference surface S_o of the shell; α_1 is meridional and α_2 is circumferential;
z is coordinate along the normal $\mathbf{n}$ to S_o;
A_1, A_2 are the coefficients of Lamé;
k_1, k_2 are the main curvature;
U, W are components of the displacement vector of an arbitrary shell point;
u, w are components point displacement vector of the surface S_o;
ϑ_1 is the rotation angle of the normal $\mathbf{n}$;
ε_{jk} are components of the strain tensor;
E_{11}, E_{22} are components of the tangential strain of tension-compression in the directions of coordinates α_1 and α_2 on S_o;
K_{11}, K_{22} are components of bending deformation (changes in the main curvatures);
$(\ldots)' = (\ldots)_{,\alpha_1}/A_1 = d(\ldots)/(A_1 d\alpha_1)$ is the differential operator.

Here, the positive direction of the normal is considered to be the external direction to the shell.

The equations of equilibrium in forces and moments correspond to kinematics (8.1.2) based on Lagrange principle

$$T'_{11} + \psi(T_{11} - T_{22}) + K_1 Q_{11} + q_1 = 0,$$

$$Q'_{11} + \psi Q_{11} - K_1 T_{11} - K_2 T_{22} + q_3 = 0,$$

$$M'_{11} + \psi(M_{11} - M_{22}) - Q_{11} - T_{11}\vartheta_1 = 0. \qquad (8.1.3)$$

A variant of linear theory follows from (8.1.2), (8.1.3), if we discard the quadratic terms. Further, we will limit ourselves to a variant of an isotropic linear elastic material for which Hooke's law holds:

$$\sigma_{11} = [E/(1 - \nu^2)](\varepsilon_{11} + \nu\varepsilon_{22}), \quad \sigma_{22} = [E/(1 - \nu^2)](\varepsilon_{22} + \nu\varepsilon_{11}). \qquad (8.1.4)$$

Taking into account the hypotheses of Kirchhoff shell theory, the stress intensity is determined by the formula

$$\sigma = [\sigma_{11}^2 + \sigma_{22}^2 - \sigma_{11}\sigma_{22}]^{1/2}. \qquad (8.1.5)$$

According to Mises' strength criterion, the material operates within the elastic limits at $\sigma \leq \sigma_{as}$, where the normally permissible stress intensity σ_{as} is equal to σ_{ys}, the yield strength.

The elasticity relations for isotropic shells have the form

$$T_{11} = B(E_{11} + \nu E_{22}), \quad T_{22} = B(E_{22} + \nu E_{11}),$$

$$M_{11} = D(K_{11} + \nu K_{22}), \quad M_{22} = D(K_{22} + \nu K_{11}); \qquad (8.1.6)$$

$$B = Eh/(1 - \nu^2), \quad D = Eh^3/[12(1 - \nu^2)],$$

where B and D are the effective tensile and flexural stiffness of the shell, E is Young's modulus, and ν is Poisson coefficient.

The boundary conditions follow from the requirement of vanishing contour integrals of Lagrange functional. In the case of uniform boundary conditions at the ends of the shell, either generalized displacements or generalized forces are equal to zero

$$u(1 - \gamma_1) + \gamma_1 T_{11} = 0, \quad w(1 - \gamma_2) + \gamma_2 Q_{11} = 0,$$

$$\vartheta_1(1 - \gamma_3) + \gamma_3 M_{11} = 0 \qquad (8.1.7)$$

when $\alpha_1 = \alpha_{1l}$(on the left edge);

$$u(1 - \gamma_4) + \gamma_4 T_{11} = 0, \quad w(1 - \gamma_5) + \gamma_5 Q_{11} = 0,$$

$$\vartheta_1(1 - \gamma_6) + \gamma_6 M_{11} = 0 \qquad (8.1.8)$$

when $\alpha_1 = \alpha_{1r}$ (on the right edge), where γ_j takes the values 0 or 1.

8.2. Conditions of Shell Sections Connection on the Discrete Rings

In the equations under consideration, the displacement components and force factors are decomposed along the axes of the trihedron of the main surface. Therefore, for composite shells formed by sections of different geometries and having meridian breaks, stiffness jumps, ring loads, and other violations of continuity of properties, the interface conditions must be met on the break lines. In this case, the sections are usually joined through annular edges [65, 66]. To form these conditions, the equilibrium equations of the ring edge are used taking into account the reactions of adjacent sections of shells.

The parameters and reactions of the left section are distinguished by the index-the sign "−", the right "+". The coordinate systems associated with the cross-section of the ring edge are shown in Fig. 8.2.1. Here, the axis $\bar{\alpha}_1 \equiv \alpha_{1r}$ is parallel to the axis of rotation of the shell; the axis $\bar{\alpha}_3 \equiv \alpha_{3r}$ has the direction of the polar radius of the cylindrical coordinate system; α_2 is counted along the

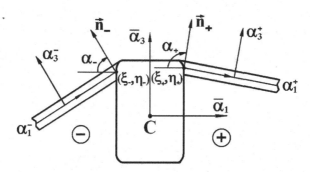

Fig. 8.2.1. Coordinate system in the transverse cross section of the rib.

centerline of the edge; and α_-, α_+ are the angles of inclination of the normal (meridian) at the junction point.

The interface conditions have the following form:

$$T_{11+} = T_{3+}\sin\alpha_+ + Q_{1+}\cos\alpha_+,\, Q_{11+} = Q_{1+}\sin\alpha_+ - T_{3+}\cos\alpha_+,$$

$$M_{11+} = [r_-y_{3-} - M_{3k} - r_k m_k + r_-(\eta_- T_{3-}$$
$$- \xi_- Q_{1-})]/r_+ - \eta_+ T_{3+} + \xi_+ Q_{1+},$$

$$u_+ = u_{1+}\cos\alpha_+ + u_{3+}\sin\alpha_+,$$

$$w_+ = u_{1+}\sin\alpha_+ - u_{3+}\cos\alpha_+,\, \vartheta_{1+} = \vartheta_{1-}, \qquad (8.2.1)$$

where

$$u_{1-} = u_-\cos\alpha_- + w_-\sin\alpha_-,\quad u_{3-} = u_-\sin\alpha_- - w_-\cos\alpha_-,$$

$$u_{1+} = u_{1-} - (\xi_+ - \xi_-)\vartheta_{1-},\quad u_{3+} = u_{3-} + (\eta_+ - \eta_-)\vartheta_{1-},$$

$$T_{3-} = T_{11-}\sin\alpha_- - Q_{11-}\cos\alpha_-,$$

$$Q_{1-} = T_{11-}\cos\alpha_- + Q_{11-}\sin\alpha_-,$$

$$u_k = u_{3-} - \eta_-\vartheta_{1-},\quad w_k = u_{1-} + \xi_-\vartheta_{1-},$$

$$\varepsilon_{2k} = w_k/r_k,\quad \chi_3 = -\vartheta_{1-}/r_k,$$

$$T_{2k} = E_k F_k \varepsilon_{2k},\quad M_{1k} = E_k J_{13k}\chi_3,\quad M_{3k} = E_k J_{3k}\chi_3,$$

$$Q_{1+} = r_- Q_{1-}/r_+ + T_{2k}/r_+ - r_k q_k/r_k,$$

$$T_{3+} = r_- T_{3-}/r_+ - r_k t_k/r_+. \qquad (8.2.2)$$

In the absence of edges, the formulas that provide the transformation of displacements and generalized internal forces on the meridian break are given the form

$$u_{3-} = u_-\sin\alpha_- - w_-\cos\alpha_-,\quad u_{1-} = u_-\cos\alpha_- + w_-\sin\alpha_-,$$

$$T_{3-} = T_{11-}\sin\alpha_- Q_{11-}\cos\alpha_-,\quad Q_{1-} = T_{11-}\cos\alpha_- + Q_{11-}\sin\alpha_-,$$

$$u_{3+} = u_{3-},\quad u_{1+} = u_{1-},\quad T_{3+} = T_{3-},$$

$$Q_{1+} = Q_{1-},\quad M_{11+} = M_{11-},$$

$$u_+ = u_{3+}\sin\alpha_+ + u_{1+}\cos\alpha_+,\quad w_+ = -u_{3+}\cos\alpha_+ + u_{1+}\sin\alpha_+,$$

$$T_{11+} = T_{3+}\sin\alpha_+ - Q_{1+}\cos\alpha_+,$$

$$Q_{11+} = -T_{3+}\cos\alpha_+ + Q_{1+}\sin\alpha_+. \qquad (8.2.3)$$

The interface conditions include the equations of equilibrium of forces and moments and formulas for converting coordinate systems in rotation. By applying these conditions at the ends, you can set edge offsets or forces in the desired direction by controlling the approach and departure angles α_- and α_+.

To apply numerical methods for solving boundary value problems, it is necessary to bring the equations to the canonical form.

Let's choose the values that are included in the natural boundary conditions as the main functions. This makes it easy to meet conditions at borders. We introduce the following unified designations for these functions:

$$y_1 = T_{11}, \quad y_2 = Q_{11}, \quad y_3 = M_{11}, \quad y_4 = u, \quad y_5 = w, \quad y_6 = \vartheta_{1-}.$$
$$(8.2.4)$$

Then, the initial relations of the theory for isotropic shells can be reduced to a system of ordinary differential equations of the sixth order in canonical form

$$\boldsymbol{y}' = F(\alpha_1, \boldsymbol{y}) + P(\alpha_1), \tag{8.2.5}$$

where

$$\boldsymbol{y} = (y_1, \ldots, y_6), \quad \boldsymbol{f} = (f_1, \ldots, f_6),$$
$$\boldsymbol{p} = (-q_1, -q_3, 0, 0, 0, 0), \tag{8.2.6}$$
$$f_1 = \psi(T_{22} - y_1) - k_1 y_2,$$
$$f_2 = -\psi y_2 + k_1 y_1 + k_2 T_{22},$$
$$f_3 = \psi(M_{22} - y_3) + y_2 + y_1 y_6,$$
$$f_4 = E_{11} - y_1 - k_1 y_5 - 0.5(y_6)^2,$$
$$f_5 = k_1 y_4 - y_6,$$
$$f_6 = K_{11}, \tag{8.2.7}$$

$$E_{22} = \psi y_4 + k_2 y_5, \quad k_{22} = \psi y_6,$$
$$K_{11} = y_3/D - \nu K_{22}, \quad E_{11} = y_1/B - \nu E_{22},$$
$$T_{22} = B(E_{22} + \nu E_{11}), \quad M_2 = D(K_{22} + \nu K_{11}). \tag{8.2.8}$$

Boundary conditions relative to the main functions can be written as

$$y_4(1 - \gamma_1) + \gamma_1 y_1 = 0, \quad y_5(1 - \gamma_2) + \gamma_2 y_2 = 0,$$
$$y_6(1 - \gamma_3) + \gamma_3 y_3 = 0, \quad \alpha_1 = \alpha_{1l}; \tag{8.2.9}$$

$$y_4(1 - \gamma_4) + \gamma_4 y_1 = 0, \quad y_5(1 - \gamma_5) + \gamma_5 y_2 = 0,$$

$$Y_6(1 - \gamma_6) + \gamma_6 y_3 = 0, \quad \alpha_1 = \alpha_{1r}. \tag{8.2.10}$$

It is advisable to implement algorithms for equations in dimensionless form. When switching to dimensionless values, the main normalizing parameters E_*, ν_*, R_*, h_*, are used. They have the meaning and dimension of characteristic quantities, respectively, Young's modulus, Poisson's ratio, radius of curvature or linear size, and thickness.

Below, dimensionless values are grouped in curly brackets with the index "U" in the left parts of the equalities. Their dimensional (normalized) analogs are implied in brackets with the index "D" in the right parts of the formulas

$$\{U, W, u, w, h, u_r, w_z, z\}_U = \{\ldots\}_D/h_*,$$

$$\{A_1, A_2, L_i\}_U = \{\ldots\}_D/R_*,$$

$$\{k_1, k_2, K_{11}, K_{22}, K_{12}\}_U = R_*\{\ldots\}_D,$$

$$\{\vartheta_1, \vartheta_2, \varepsilon_{11}, \varepsilon_{22}, E_{11}, E_{22}\}_U = \{\ldots\}_D/\varepsilon_*,$$

$$\{\sigma_{11}, \sigma_{22}, \sigma, \sigma_{as}, \sigma_{ys}\}_U = \{\ldots\}_D/(E\varepsilon_*),$$

$$\{T_{11}, T_{22}, Q_{11}, Q_{22}, T_z, Q_r\}_U = \{\ldots\}_D R_*/A_*,$$

$$\{M_{11}, M_{22}\}_U = \{\ldots\}_D R_*/D_*, \quad \{Q_1, Q_3, \}_U = \{\ldots\}_D R_*{}^2/A_*,$$

$$\{B\}_U = \{\ldots\}_D/B_*, \quad \{D\}_U = \{\ldots\}_D/D_*, \quad \{E\}_U = \{\ldots\}_D/E_*,$$

$$\{q_r, t_r\}_U = \{\ldots\}_D R_*/A_*, \quad \{m_K\}_U = \{\ldots\}_D R_*/D_*, \tag{8.2.11}$$

where

$$\varepsilon_* = h_*/R_*, \; B_* = E_* h_*/(1 - \nu_*^2), \; A_* = B_* h_*, \; D_* = A_* h_*^2. \tag{8.2.12}$$

After switching to dimensionless quantities and structuring, the system of resolving equations is reduced to the form

$$dy_i/d\alpha_1 = A_1(f_i + f_i^* + b_i), \quad i = 1, \ldots, 6; \tag{8.2.13}$$

$$f_1 = \psi(T_{22} - y_1) - k_1 y_2, \quad f_2 = -\psi y_2 + k_1 y_1 + k_2 T_{22},$$

$$f_3 = \psi(M_{22} - y_3) + y_2/\varepsilon_*, \quad f_4 = E_{11} - y_1 - k_1 y_5,$$

$$f_5 = k_1 y_4 - y_6, \quad f_6 = K_{11}/\varepsilon_*;$$

$$f_1^* = 0, \quad f_2^* = 0, \quad f_3^* = y_1 y_6,$$

$$f_4^* = -0.5\varepsilon_* y_6^2, \quad f_5^* = 0, \quad f_6^* = 0;$$

$$b_1 = -q_1, \quad b_3 = -q_3; \tag{8.2.14}$$

$$E_{22} = \psi y_4 + k_2 y_5, \quad k_{22} = \varepsilon_* \psi y_6, \quad k_{11} = y_3/D - \nu K_{22},$$

$$E_{11} = y_1/B - \nu E_{22}, \quad T_{22} = B(E_{22} + \nu E_{11}),$$

$$M_{22} = D(K_{22} + \nu K_{11}). \tag{8.2.15}$$

Here, the "U" index is omitted to simplify writing. In this case, the type of coupling conditions, the component of strain tensors ε_{ij} and stress σ_{ij}, stress intensity, and boundary conditions are preserved the same and in dimensionless form.

Let's go to (8.2.13)–(8.2.15) to completely resolve the functions. The following relations exist:

$$E_{11} = y_1/B - \nu(\psi y_4 + k_2 y_5),$$

$$T_{22} = B(1 - \nu^2)(\psi y_4 + k_2 y_5) + \nu y_1,$$

$$M_{22} = D(1 - \nu^2)\varepsilon_* \psi y_6 + \nu y_3. \tag{8.2.16}$$

Then the equations (8.2.13) get the following canonical form:

$$dy_1/d\alpha_1 = A_1\{\psi(1 - \nu)[B(1 + \nu)(\psi y_4 + k_2 y_5) - (1 - \nu)y_1]$$
$$- k_1 y_2 - q_1\},$$

$$dy_2/d\alpha_1 = A_1\{-\psi y_2 + (k_1 + \nu k_2)y_1$$
$$+ k_2 B(1 - \nu^2)(\psi y_4 + k_2 y_5) - q_3\},$$

$$dy_3/d\alpha_1 = A_1\{\psi(1 - \nu)[D(1 + \nu)\varepsilon_* \psi y_6 - y_3] + y_2/\varepsilon_* + y_1 y_6\},$$

$$dy_4/d\alpha_1 = A_1\{y_1/B - \nu\psi y_4 - (k_1 + \nu k_2)y_5 - \varepsilon_*(y_6)^2\},$$

$$dy_5/d\alpha_1 = A_1\{k_1 y_4 - y_6\},$$

$$dy_6/d\alpha_1 = A_1\{y_3/(\varepsilon_* D) - \nu\psi y_6\}. \tag{8.2.17}$$

This standard form is necessary for applying methods that reduce boundary value problems to procedures for solving Cauchy problems. To specify Lame coefficients A_1, A_2, the main curvatures k_1, k_2, and the parameter ψ, we should use knowledge from differential geometry. Math reference books can help you here. For typical rotation shells, a number of formulas are given in [65].

8.3. The Equations for the Deformations of Shells of Large Angles of Rotation

In problems for axisymmetric shells, equations are also applied that do not impose restrictions on rotation angles and take into account transverse shear deformations. These include the equations of E. Reissner [44]. The corresponding deformation relations in the main designations of this work have the form

$$\varepsilon_1 = (u' \cos \Phi + w' \sin \Phi)/\alpha_0 + \cos(\Phi - \Phi_o) - 1, \quad \varepsilon_2 = u/r_o,$$

$$\kappa_1 = (\Phi'_o - \Phi')/\alpha_o, \quad \kappa_2 = (\sin \Phi_o - \sin \Phi)/r_o,$$

$$\gamma_o = (w' \cos \Phi - u' \sin \Phi)/\alpha_o + \sin(\Phi - \Phi_o), \qquad (8.3.1)$$

where

w, u are displacements in the direction of the shell axis and the vertical axis direction;

Φ_o is the angle of inclination of the internal normal of the undeformed shell to the axis of rotation;

Φ is the angle of inclination of the internal material normal of the deformed shell to the axis of rotation;

ε_1, ε_2 are components of deformations (relative elongations in the directions of meridians and parallels);

κ_1, κ_2 are components of curvature changes;

γ_o is transverse shear strain (angle);

r_o is radius of the middle surface, measured from the axis of rotation;

α_o is Lame coefficient for meridian: $ds = \alpha_o d\xi$;

ξ is curved coordinate calculated along the meridian;

$(\ldots)' = d(\ldots)/d\xi$ is the derivative with respect to the meridional coordinate.

For components (8.3.1), there is a condition for compatibility of deformations (8.3.2) and an equation for axial displacement (8.3.3)

$$(r_o\varepsilon_2)' - \alpha_o\varepsilon_1 \cos \Phi + \alpha_o\gamma_o \sin \Phi - \alpha_o(\cos \Phi - \cos \Phi_o), \qquad (8.3.2)$$

$$w' = \alpha_o\varepsilon_1 \sin \Phi + \alpha_o\gamma_o \cos \Phi + \alpha_o \sin \Phi - \sin \Phi_o). \qquad (8.3.3)$$

The statics equations follow from Lagrange principle based on (8.3.1) in the form

$$(r_o V)' + \alpha_o r_o p_w = 0,$$

$$(r_o H)' - \alpha_o N_2 + \alpha_o r_p p_u = 0,$$

$$(r_o M_1)' - \alpha_o M_2 \cos \Phi + \alpha_o r_o Q = 0, \qquad (8.3.4)$$

where

$$V = N_1 \sin \Phi + Q \cos \Phi, \quad H = N_1 \cos \Phi - Q \sin \Phi, \qquad (8.3.5)$$

N_1, N_2 are tangential forces;
Q is cutting force;
V, H are axial and radial components of internal forces;
M_2, M_2 are bending moments, respectively, along the meridiane and in the circumferential direction.

For the considered deformations, Hooke's law of a transversely isotropic material, the stress intensity, and the shell elasticity relations have the following form:

$$\sigma_1 = [E/(1 - \nu^2)](e_1 + \nu e_2),$$

$$\sigma_2 = [E/(1 - \nu^2)](e_2 + \nu e_1), \quad \sigma_{13} = G\gamma_o, \qquad (8.3.6)$$

$$e_1 = \varepsilon_1 + z\kappa_1, \quad e_2 = \varepsilon_2 + z\kappa_2; \qquad (8.3.7)$$

$$N_1 = B(\varepsilon_1 + \nu\varepsilon_2), \quad N_2 = B(\varepsilon_2 + \nu\varepsilon_1), \quad Q = G_{13}\gamma_o,$$

$$M_1 = D(\kappa_1 + \nu\kappa_2), \quad M_2 = D(\kappa_2 + \nu\kappa_1); \qquad (8.3.8)$$

$$B = Eh/(1 - \nu^2), \quad D = Eh^3/[12(1 - \nu^2)],$$

$$G_{13} = Gh/(1 - \nu^2),$$

where

B, D, G_{13} are effective shell stiffness for tension-compression, bending and transverse shear, respectively;
E is Young's modulus, G is transverse shear modulus, and ν is Poisson's ratio.

Taking into account the hypotheses of the shell theory, the stress intensity is determined by the formula

$$\sigma = [(\sigma_1)^2 + (\sigma_2)^2 - \sigma_1\sigma_2 + 3(\sigma_{13})^2]^{1/2}. \qquad (8.3.9)$$

According to Mises' strength criterion, the material operates within the elastic limits $\sigma \leq \sigma_{as}$, where the normally permissible stress intensity σ_{as} is equal to σ_{ys} — the yield strength.

Put

$$T = r_o V, \quad \Psi = r_o H, \quad M = r_o M_1. \tag{8.3.10}$$

The value T is called the load function, and Ψ is the stress function [44].

It is possible to construct a fourth-order resolving system from two second-order nonlinear equations each with respect to the angle Φ and the stress function Ψ. However the canonical system of the sixth order is more flexible and adapted to the algorithms of methods for immersing boundary value problems in Cauchy problems

$$\boldsymbol{y}' = \boldsymbol{f}(\alpha_1, \boldsymbol{y}), \quad \boldsymbol{y} = \{y_0, \ldots, y_5\}, \quad \boldsymbol{f} = \{f_0, \ldots, f_5\} \tag{8.3.11}$$

regarding the functions

$$y_0 = T, \quad y_1 = \Psi, \quad y_2 = M, \quad y_3 = w, \quad y_4 = u, \quad y_5 = \Phi.$$

As the last main variable, it is also convenient to take $y_5 = \Phi_0 - \Phi$. It is numbered from zero to five.

The procedure for calculating the right parts of the system (8.3.36) in this case is reduced to a sequence of formulas

$$N_1 = (T \sin \Phi + \Psi \cos \Phi)/r_o, \quad Q = (T \cos \Phi - \Psi \sin \Phi)/r_o,$$

$$\varepsilon_2 = u/r_o, \quad \kappa_2 = (\sin \Phi_o - \sin \Phi)/r_o, \quad \varepsilon_1 = (N_1 - \nu B \varepsilon_2)/B,$$

$$\gamma_o = Q/G_{13}, \quad N_2 = B(\varepsilon_2 + \nu \varepsilon_1),$$

$$\kappa_1 = (M/r_o - \nu D \kappa_2)/D, \quad M_2 = D(\kappa_2 + \nu \kappa_1); \tag{8.3.12}$$

$$f_0 = -\alpha_0 r_o p_w, \quad f_1 = \alpha_o N_2 - \alpha_o r_o p_u, \quad f_2 = \alpha_o M_2 \cos \Phi - \alpha_o r_o Q,$$

$$f_3 = \alpha_o \varepsilon_1 \cos \Phi - \alpha_o \gamma_o \sin \Phi + \alpha_o (\cos \Phi - \cos \Phi_o),$$

$$f_4 = \alpha_o \varepsilon_1 \sin \Phi + \alpha_o \gamma_o \cos \Phi + \alpha_o (\sin \Phi - \sin \Phi_o),$$

$$f_5 = \Phi_o' - \alpha_o \kappa_1. \tag{8.3.13}$$

When constructing algorithms, it is advisable to express the right parts explicitly through the main functions.

For shells with two ends on the edge contours, you can set the following options for edge conditions:

(1) $w = 0$, $u = 0$, $\Phi_o - \Phi = 0$ — hard pinching;
(2) $T = Te$, $u = 0$, $\Phi_o - \Phi = 0$ — pinching, movable in the axial direction, with the specified axial force;
(3) $w = 0$, $u = 0$, $M = 0$ — fixed hinge;
(4) $w = w_e$, $M = 0$, $u = 0$ — hinge, movable in the axial direction, with a given displacement of the contour.

Implementation of the algorithm, calculations, and parametric studies should be performed in a dimensionless form. The transition to dimensionless quantities can be performed using the following formulas:

$$\{u, w, h, w_e\}_U = (m_*/h_*)\{\ldots\}_D, \quad \{r_o\}_U = \{r_o\}_D/R_*,$$

$$\{p_w, p_u\}_U = [m_* R_*^2/(E_* h_*^2)]\{p_w, p_u\}_D,$$

$$\{T, \Psi, T_e\}_U = [m_*/(E_* h_*^2)]\{\ldots\}_D,$$

$$\{N_1, N_2, Q\}_U = [m_* R_*/(E_* h_*^2)]\{\ldots\}_D,$$

$$\{M_1, M_2\{_U = [m_*^2 R_*/(E_* h_*^3)]\{\ldots\}_D,$$

$$\{\varepsilon_1, \varepsilon_2, \gamma_0\}_U = (m_*/\varepsilon_*)\{\ldots\}_D,$$

$$\{\sigma_1, \sigma_2, \sigma_{13}, \sigma, \sigma_{as}, \sigma_{ys}\}_U = [m_*/(E_* \varepsilon_*)]\{\ldots\}_D,$$

$$\{B, G_{13}\{_U = \{\ldots\}_D/(E_* \varepsilon_*), \quad \{E\{_U = \{E\}_D/E_*,$$

$$\{D\}_U = [m_*^2/(E_* h_*^3)]\{\ldots\}_D/D_*, \quad \{\kappa_1, \kappa_2\}_U = R_*\{\ldots\}_D,$$

$$\varepsilon_* = h_*/R_*, \quad \varepsilon = \varepsilon_*/m_*, \quad m_* = [12(1 - \nu^2)]^{1/2}. \quad (8.3.14)$$

To simplify the recording, we will accept the agreement to preserve the same notation in dimensionless equations as in dimensional ones. The group of equations (8.3.12), (8.3.13) in dimensionless form gets the form

$$N_1 = (T \sin \Phi + \Psi \cos \Phi)/r_o, \quad Q = (T \cos \Phi - \Psi \sin \Phi)/r_o,$$

$$\varepsilon_2 = u/r_o, \quad \kappa_2 = (\sin \Phi_0 - \sin \Phi)/r_o, \quad \varepsilon_1 = (N_1 - \nu B \varepsilon_2)/B,$$

$$\gamma_o = Q/G_{13}, \quad N_2 = B(\varepsilon_2 + \nu \varepsilon_1),$$

$$\kappa_1 = (M/r_o - \nu D \kappa_2)/D, \quad M_2 = D(\kappa_2 + \nu \kappa_1); \quad (8.3.15)$$

$$f_0 = -\alpha_0 r_0 p_w, \quad f_1 = \alpha_0 N_2 - \alpha_0 R_0 p_u,$$

$$f_2 = \alpha_0 M_2 \cos \Phi - \alpha_0 r_0 Q / \varepsilon,$$

$$f_3 = \alpha_0 \varepsilon_1 \cos \Phi - \alpha_0 \gamma_0 \sin \Phi + \alpha_0 (\cos \Phi - \cos \Phi_0)/\varepsilon,$$

$$f_4 = \alpha_0 \varepsilon_1 \sin \Phi + \alpha_0 \gamma_0 \cos \Phi + \alpha_0 (\sin \Phi - \sin \Phi_0)/\varepsilon,$$

$$f_5 = \Phi_0' - \alpha_0 \kappa_1. \tag{8.3.16}$$

The stress intensity in dimensionless form is determined by the formulas (8.3.9), (8.3.6), where

$$e_1 = \varepsilon_1 + m_* z \kappa_1, \quad e_2 = \varepsilon_2 + m_* z \kappa_2.$$

In the case of a shell of constant thickness, $h_* = h$, $-0.5 \le z \le 0.5$. It is sufficient to calculate the intensity at the values $z = -0.5$, 0, $+0.5$, since the maximum σ by z is usually reached on the front surfaces of the shell.

The equations in Section 8.2 use a cylindrical coordinate system. Therefore, if there are meridian break points, you can do without conjugation conditions, since the resolving functions remain continuous. The ability to integrate the boundary value problem over the entire definition interval without stopping to meet the interface conditions significantly simplifies the calculation technology. In particular, when using integrated mathematical packages, you can, in some cases, do with built-in procedures without time-consuming programming of algorithms.

8.4. Quadratic-Nonlinear Equations in Cylindrical Coordinate System

The same effect can be obtained by converting the equations in Section 8.1. To do this, we need to move from the local system of the mobile accompanying trihedron to the stationary cylindrical coordinate system. The corresponding conclusion of such a resolving system of equations is made in [67].

This approach can also be applied to other, more complex, equations of non-axisymmetric deformations of composite structurally anisotropic shells of rotation in problems of statics, stability, and natural and forced vibrations.

Consider the coupling conditions (8.2.11). They include equilibrium equations and formulas for converting components in the rotation of coordinate systems. In this case, from the accompanying trihedra of sections, the transition is made to the basis of the cylindrical coordinate system. The components of displacement and internal forces are equal due to the continuity of displacement and equilibrium conditions in this system. If these components are accepted as the main ones and go to them in equations (8.1.1)–(8.1.3), then there is no need to fulfill the conjugation conditions (8.2.11). This significantly simplifies algorithms for methods of immersion of boundary value problems in Cauchy problems (methods of run-through and targeting) [65, 66], since it allows performing the process of integration of Cauchy problems without interruptions over the entire interval of shell definition. In this case, the geometric parameters of the shell that have discontinuities of the first kind (normal angles, main curvatures, etc.) must also be determined over the entire interval.

This is easily done using logical operators available in integrated math packages. For example, the angle of the meridian is determined by the superposition of piecewise functions $\alpha^i(\alpha_1)$ defined on the section with the number i and vanishing outside it

$$\beta(\alpha_1) = \sum_i \alpha^i(\alpha_1). \tag{8.4.1}$$

We introduce a set of basic functions in unified notation

$$y_1 = T_z, \ y_2 = Q_r, \ y_3 = M_{11}, \ y_4 = u_z, \ y_5 = u_r, \ y_6 = \vartheta_1. \tag{8.4.2}$$

We assume that the replacement of type (8.2.11) is performed in the entire scope of the shell definition. Here the angles of the meridian slope are denoted by β. Then,

$$u = Y_4 \sin \beta + Y_5 \cos \beta, \quad w = -Y_4 \cos \beta + Y_5 \sin \beta,$$

$$T_{11} = Y_1 \sin \beta + Y_2 \cos \beta, \quad Q_{11} = -Y_1 \cos \beta + Y_2 \sin \beta. \tag{8.4.3}$$

After substituting (8.4.3) in (8.1.1)–(8.1.3), transformations using the relations

$$A_2 = r_o, \quad \psi = \cos \beta / r_o, \quad k_1 = \beta', \quad k_2 = \sin \beta / r_o,$$

$$\psi \cos \beta + k_2 \sin \beta = 1/r_o, \quad \psi \sin \beta - k_2 \cos \beta = 0 \tag{8.4.4}$$

and resolutions relative to the derivatives of the main functions, we arrive at the system

$$Y_1' = -Y_1 \cos \beta / r_o - q_z,$$

$$Y_2' = T_{22}/r_o - Y_2 \cos \beta / r_o - q_r,$$

$$Y_3' = (M_{22} - Y_3) \cos \beta / r_o - Y_1 \cos \beta + Y_2 \sin \beta$$
$$\qquad + Y_6 (Y_1 \sin \beta + Y_2 \cos \beta),$$

$$Y_4' = E_{11} \sin \beta + Y_6 \cos \beta - 0.5(Y_6)^2 \sin \beta,$$

$$Y_5' = E_{11} \cos \beta - Y_6 \sin \beta - 0.5(Y_6)^2 \cos \beta,$$

$$Y_6' = K_{11}. \qquad (8.4.5)$$

There,

$$q_z = q_1 \sin \beta - q_3 \cos \beta, \quad q_r = q_1 \cos \beta + q_3 \sin \beta,$$

$$T_{22} = \nu(Y_1 \sin \beta + Y_2 \cos \beta) + BY_5/r_o,$$

$$M_{22} = \nu Y_3 + DY_6 \cos \beta / r_o,$$

$$E_{11} = (Y_1 \sin \beta + Y_2 \cos \beta)/B - \nu Y_5/r_o,$$

$$K_{11} = Y_3/D - \nu Y_6 \cos \beta / r_o, \qquad (8.4.6)$$

$r_o = r_o(\alpha_1)$ is the polar radius of the shell.

Then the transition to dimensionless quantities is performed according to the formulas (8.2.11). To simplify writing for dimensionless quantities, we will leave the original notation. In this case, the form of the components of deformations ε_{ii} and stresses σ_{ii}, as well as the intensity of stresses, remain formally unchanged and in a dimensionless form. The right parts (8.4.5) are given in full to the terms of the main functions.

As a result, the system of resolving equations takes the form

$$Y_1' = -Y_1 \cos \beta / r_o - q_z,$$

$$Y_2' = \nu Y_1 \sin \beta / r_o - (1 - \nu) Y_2 \cos \beta / r_o + BY_5 \cos \beta / r_o^2 - q_r,$$

$$Y_3' = -(1-\nu)Y_3 \cos\beta/r_o + \varepsilon_* DY_6/r_o^2 - (Y_1 \cos\beta - Y_2 \sin\beta)/\varepsilon_*$$
$$+ k_N Y_6 (Y_1 \sin\beta + Y_2 \cos\beta),$$
$$Y_4' = (Y_1 \sin\beta + Y_2 \cos\beta) \sin\beta/B - \nu Y_5 \sin\beta/r_o$$
$$+ Y_6 \cos\beta - k_N \varepsilon_* Y_6^2 \sin\beta/2,$$
$$Y_5' = (Y_1 \sin\beta + Y_2 \cos\beta) \cos\beta/B - \nu Y_5 \cos\beta/r_o$$
$$+ Y_6 \sin\beta - k_N \varepsilon_* Y_6^2 \cos\beta/2,$$
$$Y_6' = Y_3/(\varepsilon_* D) - \nu Y_6 \cos\beta/r_o, \qquad (8.4.7)$$

where the coefficient of nonlinearity k_N is introduced. When $k_N = 0$, the nonlinear terms are reset to zero, and the equations are converted to linear ones.

Boundary conditions can be set in a homogeneous form of the type (8.1.7), (8.1.8), relative to the one in Section 8.2, either using substitution of (8.2.3), or in accordance with the directions in which offset edges of the shell are allowed.

8.5. Calculation of Hydraulic Sealing Shells Connections

To illustrate, we apply the equations of E. Reissner equation to the calculation of the stress state of the shells of hydraulic connections.

Hydraulic transmissions are widely used in various machines due to a number of advantages (for example, compactness). However, hydraulics have a compaction problem due to the complexity of sealing connections on the pressure transfer path.

One of the methods of compaction is the force contact of spherical and conical surfaces. A fitting with an internal conical surface and a conical thread is welded to the end of one pipe. A nipple with an outer spherical surface is soldered to the end of the other pipe. They are tightened with a cap nut. The disadvantage of this design and manufacturing technology is the high cost, harmful production process, and the presence of scale, which is difficult to remove from the inner surface of the pipeline. During the operation of the connection under vibration conditions, the scale is separated and gets into the responsible transmission parts. In addition, vibration leads to cracks in the junction and depressurization of the joint.

There are alternative methods for sealing high-pressure pipe connections. One of the best designs is using an embedding ring device as a sealing element.

The known embedding rings are hollow lathes containing mainly a rigid part with a thickening, bounded by a cone and facing the cap nut. There is a thin-walled zone (about 20% of the ring length) with a cut-in edge facing the fitting. The thin-walled zone is deformed when screwing the fitting and nut; the edge is embedded in the outer surface of the pipeline, which ensures tightness of the connection. However, with repeated assembly and disassembly of the connection, as well as with frequent pressure drops, a short cut-in zone with a sharp transition to a hard thickened section can quickly exhaust the working resource due to fatigue stresses. This leads to depressurization of the responsible connection.

An attempt to master the production of such rings at one of the factories ended in failure. They were expensive, and tests showed the unreliability of rings of their own production. It became obvious that special equipment was required for their manufacture.

It was possible to solve the problem of reducing the cost of mortise rings and increase their reliability by using stamping technology, since their design is essentially short shells of rotation with a rather complex meridian shape. The selection of a rational configuration was carried out by an experimental method. The technical solution was found at the level of the invention and a patent was issued (author H.K. Kaderov). Technical conditions for ER-8, ER-12, and ER-20 embedding rings were developed, tested and put into practice.

However, it was advisable to apply mathematical modeling of the structure to optimize the shape and size associated with the stamping technology further.

Below is a preliminary calculation of the stress–strain state of embedding rings made in the form of rotation shells. Both the solving equations of the quadratic-nonlinear shell theory and the equations of E. Reissner type were used for modeling. The latter do not impose restrictions on rotation angles along the meridian and take into account transverse shear deformations.

We analyzed the possibility of applying linear equations that follow from the presented after discarding nonlinear terms to solve the problem. The rational model was chosen based on numerical experiments.

The shells under consideration are of the composite class, since they include sections of different geometries and have breaks in curvature and a meridian break. It is almost impossible to obtain analytical solutions for such shells, especially in nonlinear problems. Therefore, it is necessary to solve the problem using numerical methods and algorithms implemented in the application programs.

When you set the shape, the source data is the overall dimensions and curvatures specified for the sections. The middle surface is formed by sections of the surfaces of the cylinder, cone, and circular torus, for which conjugation problems are solved. The geometric parameters of the composite shell are included in the information for the mathematical model and algorithm for the subsequent calculation of the stress–strain state.

The calculations of the stress–strain state of shells that model hydraulic joint seals are presented below. The linearized equations of E. Reissner and the method of targeting are used. The axial loads at which the material begins to work beyond the elastic limits in local zones are estimated.

The design of the connection is shown in Fig. 8.5.1.

The designations of the main forming parameters of the sealing shell (embedding ring) are given in Fig. 8.5.2, and their average values for products ER-8, ER-12, ER-20 are in Table 8.5.1 (in *mm*).

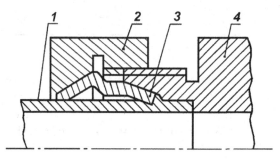

Fig. 8.5.1. Connection of pipelines with an embedding ring: 1 — pipeline, 2 — cap nut, 3 — ring embedding, 4 — fitting.

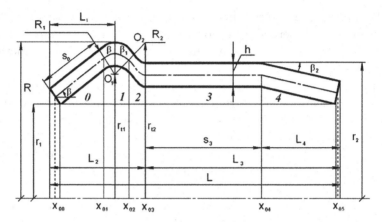

Fig. 8.5.2.   Geometric model of the embedding ring.

Table 8.5.1.   The main shaping parameters of the embedding ring.

No	$T_{ип}$	d_1	d_2	D	L	L_1	L_2	R_1	R_2	h
1	ER-8	8	9.7	10.6	9.5	2	3	1.2	0.5	0.5
2	ER-12	12	14.2	15.4	10	2.7	4	1.6	0.8	0.8
3	ER-20	20	22.4	24.5	11	3.5	5	2	1	1

Table 8.5.2.   Matrix of normalized parameters.

k_0	k_1	k_2	k_3	k_4	k_5	k_6	k_7	k_8
1	1.212	1.325	1.875	0.25	0.375	0.15	0.062	0.062
1	1.183	1.283	0.833	0.225	0.333	0.133	0.067	0.067
1	1.12	1.225	0.55	0.175	0.25	0.1	0.05	0.05

The shell that models the structure is made up of 5 elements with numbers from 0 to 4: 0 — initial cone; 1 — circular torus 1; 2 — circular torus 2, 3 — cylinder, 4 — final cone. After normalizing the data in Table 8.5.1 with the value d_1, we get the Table 8.5.2 (matrix).

Denote: $r_1 = d_1/2$, $r_2 = d_2/2$, $R = D/2$, $h_1 = k_8$. Let's take the values as normalizing values $R_* = r_1$, $h_* = h$ and let's move on to dimensionless parameters.

$$(d_1, d_2, r_1, r_2, D, R, L, L_1, L_2, R_1, R_2)_U = (\ldots)_D/R_*,$$

$$\{h\}_U = \{h\}_D/h_*. \qquad (8.5.1)$$

Below, only dimensionless parameters are used, while preserving the corresponding dimensional notation (in other words, the "U" index is omitted to simplify writing).

In accordance with Table 8.5.2,

$$r_1 = k_0, \quad r_2 = k_1, \quad R = k_2, \quad L = 2k_3, \quad L_1 = 2k_4, \quad L_2 = 2k_5,$$

$$R_1 = 2k_6, \quad R_1 = 2k_6, \quad h_1 = k_8. \qquad (8.5.2)$$

To parameterize the main surface of the shell of a composite geometry, you must find the angles of taper, the coordinates of the interface points of sections, and the lengths of the meridian arcs of these sections. As a function of an independent variable, coordinates calculated by the meridian or axis of rotation, the dependencies of the polar radius, the angles of inclination of the normal to the centerline, and derivatives must be determined.

In accordance with Fig. 8.5.2,

$$\lambda = R_1 \sin \beta, \quad \lambda_1 = L_1 - \lambda, \quad s_0 = \lambda_1 \cos \beta,$$

$$r_{00} = r_1 + 0.5h_1 \cos \beta, \quad r_{03} = r_2 - 0.5h_1, \quad r_{04} = r_{03},$$

$$H_1 = R_1(1 - \cos \beta), \quad H_2 = R_2(1 - \cos \beta); \qquad (8.5.3)$$

$$R = r_1 + h_1 \cos \beta + s_0 \sin \beta. \qquad (8.5.4)$$

From (8.5.4), a trigonometric equation for determining β follows

$$r_1 + h_1 + (L_1 - R_1 \sin \beta) + R_1(1 - \cos \beta)\mathrm{tg}\beta - R = 0. \qquad (8.5.5)$$

The parameters that depend on it are recalculated based on the found value β (8.5.3). For the angle β_1, the following relations take place:

$$R - r_2 - (R_1 + R_2)(1 - \cos \beta_1) = 0,$$

$$\beta_1 = \arccos[1 - (R - r_2)/(R_1 + R_2)]. \qquad (8.5.6)$$

Further values are found

$$r_{t1} = R - R_1, \quad R_{1c} = R_1 - 0.5h_1, \quad r_{t2} = r_2 + R_2,$$

$$R_{2c} = R_2 + 0.5h_1,$$

$$L_2 = R_1 \sin \beta + (R_1 + R_2) \sin \beta_1 + \lambda_1. \qquad (8.5.7)$$

The angle β_2 is found as the root of the equation

$$r - r_1 - 0.5h_1(1 + \cos\beta_2) - (1 - k)(L - L_2 - 0.5h_1 \sin\beta_2) = 0.$$

$$(8.5.8)$$

We introduce a variable coefficient of proportion k, $0 < k < 1$, through which we determine the ratio between s_3 and L_4: $s_3 = kL_3$, $L_4 = (1 - k)L_3$.

For L_3 and s_4, we have

$$L_3 = L - L_2 - 0.5h_1 \sin\beta_2, \quad s_4 = (L_3 - s_3)/\cos\beta_2. \quad (8.5.9)$$

Let's consider a variant of counting the longitudinal coordinate on the axis of rotation. Then the values of the coordinates of the start and end points of the entire interval and the points of intersection of the sections will be as follows:

$$x_{00} = 0, \quad x_{01} = x + s\cos\beta, \quad x_{02} = x_{01} + R_{1c}(\sin\beta - \sin\beta_1),$$

$$x_{03} = x_{02} + R_{2c}\sin\beta_1, \quad x_{04} = x_{03} + s_3, \quad x_{05} = x_{04} + s_4\cos\beta_2.$$

$$(8.5.10)$$

The current radii of the median surface in the cylindrical coordinate system for sections are determined by the functions

$$p_0(x) = r_{00} + (x - x_{00})\text{tg}\beta, \quad p_1(x) = r_{t1} + [(R_{1c})^2 - (x - x_{t1})^2]^{1/2},$$

$$p_2(x) = r_{t2} - [(R_{2c})^2 - (x - x_{t2})^2]^{1/2}, \quad p_3(x) = r_{03},$$

$$p_4(x) = r_{04} - (x - x_{04})\text{tg}\beta_2. \quad (8.5.11)$$

Let's introduce locally defined functions

$$P_j(x) = \begin{cases} p_j(x), & x \in [x_{0j}; x_{0,j+1}], \\ 0, & x \notin [x_{0j}; x_{0j,j+1}], \quad j = 0, \ldots, 4. \end{cases} \quad (8.5.12)$$

Then over the entire length of the shell, the current radius of the median surface is determined by the superposition

$$r_o(x) = P(x) = \sum_{j=0}^{4} P_j(x). \quad (8.5.13)$$

Similarly, other parameterization functions of the median surface that are included in the equations of the mathematical model for determining the stress–strain state are set over the entire interval.

Table 8.5.3.   Angle values.

	β	β_1	β_2
ER-8	26.4	42.7	6.4
ER-12	21.8	47.2	8
ER-20	23.3	53.1	5.8

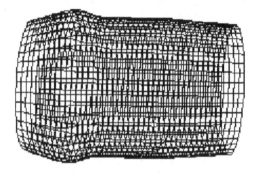

Fig. 8.5.3.   The middle surface of the ring ER-8.

The calculations of angles β, β_1 и β_2 in degrees at $k = 0.6$ for the three variants under consideration are presented in Table 8.5.3. The grid view of the middle surface of the ER-8 variant is shown in Fig. 8.5.3.

To calculate the stress–strain state, we use the dimensionless equations of E. Reissner (8.4.5), (8.4.6), converting them into the canonical form of a differential system of equations of the first order

$$y' = f(s, y), \quad y = \{y_0, \ldots, y_5\}, \quad f = \{f_0, \ldots, f_5\}. \qquad (8.5.14)$$

Let us take as the main (resolving) functions

$$y_0 = T, \quad y_1 = \Psi, \quad y_2 = M, \quad y_3 = w, \quad y_4 = u, \quad y_5 = \Phi_o - \Phi. \qquad (8.5.15)$$

We transform the right parts of the canonical system into a closed form, expressing them completely through resolving functions. In this form, they are suitable for applying the "targeting" procedures of the integrated MathCad package. In this case, the right parts for the isotropic shell in the absence of surface load will have the following form:

$$f_0(s) = 0,$$

$$f_1(s) = [r_o(s)]^{-1}\{\nu[y_0\sin(\Phi_o(s) - y_5) + y_1\cos(\Phi_o(s) - y_5)]$$
$$+ B(1 - \nu^2)y_4\},$$

$$f_2(s) = [r_o(s)]^{-1}\{\nu y_2 + D(1 - \nu^2)[\sin(\Phi_o(s)) - \sin(\Phi_o(s) - y_5)]\}$$
$$\times \cos(\Phi_o(s) - y_5) - \varepsilon^{-1}[y_o\cos(\Phi_o(s) - y_5)$$
$$- y_1\sin(\Phi_o(s) - y_5)],$$

$$f_3(s) = [r_o(s)]^{-1}\{B^{-1}[y_0\sin(\Phi_o(s) - y_5) + y_1\cos(\Phi_o(s) - y_5)]$$
$$- \nu B y_4\}$$
$$\times \sin(\Phi_o(s) - y_5) + G^{-1}[y_0\cos(\Phi_o(s) - y_5)$$
$$- y_1\sin(\Phi_o(s) - y_5)]$$
$$\times \cos(\Phi_o(s) - y_5) + \varepsilon^{-1}[\sin(\Phi_o(s) - y_5) - \sin(\Phi_o(s))],$$

$$f_4(s) = [r_o(s)]^{-1}\{B^{-1}[y_0\sin(\Phi_o(s) - y_5) + y_1\cos(\Phi_o(s) - y_5)]$$
$$- \nu B y_4\}$$
$$\times \cos(\Phi_o(s) - y_5) - G^{-1}[y_0\cos(\Phi_o(s) - y_5)$$
$$- y_1\sin(\Phi_o(s) - y_5)]$$
$$\times \sin(\Phi_o(s) - y_5) + \varepsilon^{-1}[\cos(\Phi_o(s) - y_5) - \cos(\Phi_o(s))],$$

$$f_5(s) = [r_o(s)]^{-1}\{D^{-1}y_2 - \nu[\sin(\Phi_o(s) - y_5) - \sin(\Phi_o(s))]\}, \quad (8.5.16)$$

where $r_o(s)$ is the polar radius of the median surface of the shell. The independent coordinate here is the length of the meridian arc s, measured from the left end of the shell. There is no surface load. Loading is performed by axial compression at the ends.

The equations are written in dimensionless form, the transition to which is made according to the formulas (8.2.11). Linear approximation was sufficient. We linearize the equations (8.5.16) with respect to the rotation angle y_5. The right parts of the corresponding equations have the form

$$f_0(s) = 0,$$

$$f_1(s) = [r_o(s)]^{-1}\{\nu[y_0\sin(\Phi_o(s) - y_5) + y_1\cos(\Phi_o(s) - y_5)] + B(1 - \nu^2)y_4\},$$

$$f_2(s) = [r_o(s)]^{-1}\{[\nu y_2 + D(1 - \nu^2)[\sin(\Phi_0(s)) - \sin(\Phi_0(s) - y_5)]\}$$
$$\times \cos(\Phi_o(s) - y_5) - \varepsilon^{-1}[y_0 \cos(\Phi_o(s) - y_5) - y_1 \sin(\Phi_0(s) - y_5)],$$

$$f_3(s) = [r_o(s)]^{-1}\{B^{-1}[y_0 \sin(\Phi_o(s) - y_5) + y_1 \cos(\Phi_o(s) - y_5)] - \nu B y_4\}$$
$$\times \sin(\Phi_o(s) - y_5) + G^{-1}[y_0 \cos(\Phi_o(s) - y_5) - y_1 \sin(\Phi_o(s) - y_5)]$$
$$\times \cos(\Phi_o(s) - y_5) + \varepsilon^{-1}[\sin(\Phi_0(s) - y_5) - \sin(\Phi_o(s))],$$

$$f_4(s) = [r_o(s)]^{-1}\{B^{-1}[y_0 \sin(\Phi_o(s) - y_5) + y_1 \cos(\Phi_o(s))] - \nu B y_4\}$$
$$\times \cos(\Phi_o(s) - y_5) - G^{-1}[y_0 \cos(\Phi_o(s) - y_5)$$
$$- y_1 \sin(\Phi_o(s) - y_5)] \times \sin(\Phi_o(s) - y_5)$$
$$+ \varepsilon^{-1}[\cos(\Phi_o(s) - y_5) - \cos(\Phi_0(s))],$$

$$f_5(s) = [r_o(s)]^{-1}\{D^{-1}y_2 - \nu[\sin(\Phi_0(s) - y_5) - \sin(\Phi_o(s))]\}. \quad (8.5.17)$$

We will consider a construction with the boundary conditions of a fixed hinge on the right edge and an axially movable hinge on the left edge with a given axial force T_e.

Consider the behavior of the stress intensity for the ER-8 variant, presented in dimensionless form in Fig. 8.5.4.

Here, the solid curve corresponds to the stress intensity on the middle surface $z = 0$, the dash-dotted line σ on the outer front surface

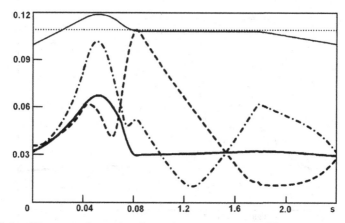

Fig. 8.5.4. The stress intensity along the meridian ring ER-8 depending on the dimensionless length of the meridian arc.

$z = -0.5$, the dashed line σ on the inner surface $z = +0.5$ of the shell. The permissible level (dimensionless σ_{ys}) is set by the constant 0.109 (horizontal line of points). The upper solid curve corresponds to a graph of the meridian shape with a scaling factor, which allows you to see the intensity of stresses at characteristic points.

The area of torus shells is most stressed. Peak stress levels occur in the vicinity of the interface zones with the torus. The first plastic deformations occur on the inner surface in the interface zone of the second torus element with the cylinder.

Here, the solid curve corresponds to the intensity of stresses σ on the middle surface $z = 0$, the dash-dotted line σ on the outer face surface $z = -0.5$, and the dashed line σ on the inner surface $z = +0.5$ of the shell. The acceptable level (dimensionless σ_{ys}) is set by the constant 0.109 (horizontal line of points).

For the ER-12 variant (Fig. 8.5.5), the first plastic deformations can appear almost simultaneously both on the inner surface in the interface zone of the second torus with the cylinder and on the outer surface of the first torus at the highest value of the polar radius.

In the ER-20 variant (Fig. 8.5.6), the picture is similar to ER-8, and the main intensity peaks differ slightly.

In the case of a 40X steel material with Young's modulus $E = 218.5$ GPa and a yield strength $\sigma_{ys} = 900\,MPa$, the structures work elastically up to the values of the T_e axial force (in Newtons):

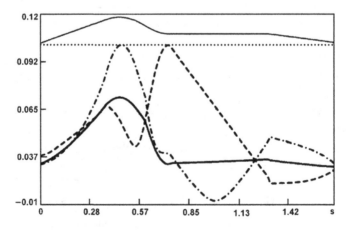

Fig. 8.5.5.   Stress intensity along the ER-12 meridian.

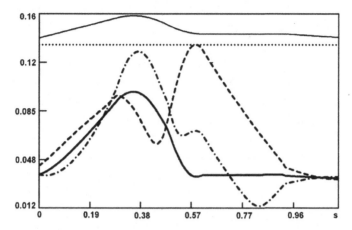

Fig. 8.5.6.   Stress intensity along the ER-20 meridian.

for the variant ER-8, 3639 N; for ER-12, 10009 N; and for ER-20, 16110 N.

Further development of modeling of these structures of sealing shells is associated with the solution of rather complex contact problems. In this case, it is necessary to take into account the work of the material in the area of plastic deformations. In the process of force and kinematic action, the shell contacts the inner surfaces of the cap nut and the fitting (Fig. 8.5.1).

It is also important to build a model that takes into account the process of embedding the ring edge in the material of the hydraulic tube. Perhaps models from the field of metal cutting theory can help here, if they exist and are suitable for this system.

Conclusion

The process of designing new products of modern technology, buildings, and structures is currently difficult to imagine without the use of computer technology and related software. Many opportunities are possessed by complexes based on the methods of finite element analysis. However, it is known that the creation of a multifunctional computing complex requires the efforts of many specialists in the subject areas of science, mathematics, and programming. The elaboration of such complexes requires the inclusion of debugging and error diagnosis modes in the software packages to enable the user to follow in a step-by-step manner the behavior of the calculation model and, if necessary, make adjustments to the program.

The use of modern computer technologies in training systems, research, and design work in various fields is also a prerequisite for ensuring advanced level and competitiveness. According to world-leading analysts, the main success factors in modern industrial production are the reduction in the time to market for products, lower costs, and improved quality.

Effective technologies that meet these requirements include CAD/CAM/CAE-systems (hereinafter referred to as C-systems) for computer-aided design, technological preparation of production, and engineering analysis. Progress in the development of such systems has led to the creation of a market for software products produced by powerful corporations [57]. Currently, it is possible to choose and purchase both stand-alone and integrated application packages — from drawing-oriented to modeling products throughout the entire life cycle. It is clear that such systems are very expensive, time-consuming, and difficult to develop, and therefore quite conservative. They are focused mainly on use in large industrial and design organizations. The decision to purchase them refers to the level of strategic choice for investment and long-term development planning of an institution, company, and corporation.

Along with industrial packages, integrated packages (IP), which are more flexible in application, combining the capabilities of symbolic and computational mathematics, programming tools, graphics and animation, are also actively developing. Such packages are more accessible for an individual user, have a convenient interface, and have several hundreds and even thousands of built-in functions and operators. The built-in capabilities of integrated packages allow you to work effectively with differential equations and implement a variety of numerical methods. At the same time, IP can be used as a platform for algorithmic programming and solving very complex problems of strength, dynamic, and other calculations. In order to teach, IP allows you to clearly demonstrate the mathematics that is taught at school and universities, and almost in the usual recording of a text editor or paper edition. Only the pages in the IP are interactive, and the formulas come to life. Particularly user-friendly interface in this regard are the packages of the MathCad series, which are used to implement the methods of this book. It allows you to create programs that are understandable, easily modifiable, and shared with other users. Therefore, IP facilitates the exchange of information between developers of collective projects and can serve as a means of communication between the teacher and the student. This opens up new possibilities for the block implementation of programs for educational, scientific, and research purposes and the creation of libraries in the languages of integrated packages. Since they also contain multimedia and network capabilities with access to the Internet, IPs are potentially adapted to open and distance education systems. They create training courses and manuals (such as [58]). Integrated packages can be a means of learning the methods of structural engineering analysis [59] and a good addition to the development of industrial CAD and CAE systems.

On the platform of the licensed IP MathCad 2001 [60], we have developed a number of programs that can also be run on later versions of the IP. They allow you to solve effectively linear and nonlinear boundary value problems of mathematical models leading to systems of differential equations. The programs use differential sweep algorithms in combination with orthogonalization methods according to S.K. Godunov, which ensure stability of calculations over long intervals; Runge–Kutta type methods for integrating Cauchy problems; two-sided shooting method for nonlinear boundary value problems; modules for parameterization, construction, and visualization of complex surfaces that form the appearance of complex shell structures; and many others. The need to program some typical modules is due to the fact that Mathcad's built-in procedures for solving boundary and initial problems work only in the field of real numbers. Programmed modules can work with equations in complex form. This is important for solving vibrational dynamic problems that take into account the scattering of mechanical energy in structures through the task of complex stiffnesses. The created base is used to carry out coursework and graduation projects, research, and applied work. The problems of calculating large nonlinear deformations of structurally orthotropic shells were solved; nonlinear static problems for shells of composite geometry; problems of oscillations of statically stressed composite shell-containers containing liquid; problems of stability, vibration dynamics, damping of vibrations of reinforced shells; strength problems of sealing elements of hydraulic circuits, design of gas cylinders; and others. The accompanying electronic knowledge base has also been accumulated.

Integrated MathCad packages were developed by MathSoft (USA). Their wide and deserved popularity for the automation of mathematical calculations is due to the following reasons.

Prior to their creation, one could only dream of a mathematically oriented programming language for writing algorithms for solving mathematical, scientific, and technical problems in the most convenient, compact, and understandable form. High-level programming languages — Fortran, Algol, BASIC, Pascal — which were used for this purpose by mathematicians, physicists, and engineers and did not closely meet these requirements. Until now, MathCad packages remained the only mathematical systems in which the description of the solution of mathematical problems is given using the usual mathematical formulas and signs, and the results of the calculations can be presented in any desired form. Moreover, using them is extremely simple. Since its inception, MathCad class systems have had a convenient user interface. Most program management actions are intuitive. Despite the computing power embedded in the system, it takes quite a short time to master the basic capabilities of a person working in a Windows environment. In a word, MathCad systems are focused on the mass user.

MathCad is a mathematically oriented universal systems. A distinctive feature of MathCad is the ability to work with documents. Documents combine the description of a mathematical algorithm for solving a problem (or a series of tasks) with text comments and calculation results given in the form of symbols, numbers, tables, or graphs.

In fact, MathCad integrates three editors: formulaic, textual, and graphic. In addition to the actual calculations, they allow you to qualitatively prepare texts of articles, books, dissertations, scientific reports, diplomas, and course projects, and they also facilitate the collection of the most complex mathematical formulas and enable the presentation of results, in exquisite graphic form. Using hyperlinks, you can prepare high-quality electronic textbooks and books with many hyperlinks, high-quality texts with various emphasis, mathematical formulas, and graphs in the MathCad environment. It is important to note that such textbooks are "live": all the examples in them work and can be used with various input data set by students. This allows you to create excellent training programs for any course based on a mathematical apparatus.

The input language in MathCad is an interpretive type. This means that when it recognizes any object in the system, it immediately performs the operations indicated in the block. The system implementation language is C++.

Essentially, the input language of the system is an intermediate link between the language of communication of a document hidden from the user and the language of the implementation of the system. As the user creates (by means of text, formula, and graphic editors) objects in the editing window (texts, formulas, tables, and graphs), the system itself creates a program in some intermediate communication language, which is stored in RAM until it is saved on disk as an mcd file. However, it is important to emphasize that the user is not required to know the programming languages (implementations and communications), it is enough to master the input language of the system close to the natural mathematical language. Even knowledge of the input language is reduced. Almost all operators that look like familiar mathematical symbols can be entered with the mouse. The preparation of computing units is facilitated by the derivation of the template when specifying one or another operator. For this, in Math-Cad are typesetting panels with templates of various mathematical symbols.

The first versions of the system made it possible to implement only linear programs based on the concept of a function. The If function and ranked variables in some cases could replace conditional expressions and loops, but with serious limitations. There was no possibility of setting completed software modules. Starting with the version of MathCad PLUS 6.0 PRO, the most important programming tools are included in the systems. They are concentrated in a panel of program elements. It is possible to set program blocks — procedures with generally accepted programming operators. Together with other means of the input language, this makes it unusually flexible, powerful, and visual. In modern versions of MathCad, a very elegant function has appeared for recording built-in program modules in a document that implement typical control structures. A software module in the MathCad system is marked in bold with a vertical line in the text of the document. Such blocks allow you to use all the means of not only the mathematically oriented input language MathCad but also classical programming.

Many computational methods can be solved in the MathCad system, even without using explicit software tools. However, these tools facilitate the solution of complex problems, especially when there is a description of their software implementation in any programming language. Then it is easy to translate the implementation of the solution of the problem from this language into the programming language of the MathCad system.

As noted, the input language of the MathCad system is interpretive. In the interpreters, the listing of the user program is viewed from top to bottom (and within the line from left to right), and any instructions in the program are immediately followed. It is believed that interpreters act slowly, but with modern computer power this is almost imperceptible. And although the system has introduced the possibility of expanding it with functions that are specified by ordinary programs in C or C++, you can write a program in MathCad many times faster. Moreover, any numerical method looks much more transparent. In general, it should be noted that the problem of including program blocks in documents in MathCad is solved elegantly and beautifully. It is very convenient to use the MathCad software apparatus for teaching the basics of algorithmic programming.

References

[1] Schmoeckel, D. and Luo, S., Darstellung der anisotropen induczierten plastischen Eigenschaften gewalzter Bleche, washing der Ict–the Theorie. *Bleh the Creator of the Profile*, 1992, 39(4), pp. 324–330. [in German]

[2] Annin, B.D. and Zhigalkin, V.M., *Behavior of materials under complex loading*. Novosibirsk: Publishing House of SB RAS, 1999, p. 342. [in Russian]

[3] Aral, Y., Kobayashi, H., and Tamura, M., Elastic-plastic analysis of thermal stresses for optimal material design of functionally graded material. *Trans. Japanese. Soc. Mechanic. Ang. A.*, 1993, 59(559), pp. 849–855.

[4] Romanov, K.I. *Mechanics of hot forming of metals*. M.: Mashinostroenie, 1993, p. 240. [in Russian]

[5] Kosarchuk, V.V. and Melnikov, S.A. Deformation theory of plasticity of anisotropic materials and its experimental substantiation. Proc. of 18 Science. conf. young scientists of the Institute of Mech. AS of Ukraine. Kiev, 18–21 May, 1993. Part 1, Inst Mech. Academy of Sciences of Ukraine. Kiev, 1993, pp. 53–57. Dep. in GNTB Ukraine 16.08.93, 1764–UK 93. [in Russian]

[6] Zuev, L.B. Plastic flow of solids as an autowave process in a nonlinear medium. Nonlinear dynamic analysis: 2nd Intern. Congress. Moscow, 3–8 June, 2002. M.: Publ. MAI, 2002, p. 149. [in Russian]

[7] Tomilov, M.F., Tolstov, C.A., and Tomilov, F.H. Anisotropy limit of plasticity of sheet materials. System. probl. of math. modeling quality, information, electronic and laser technologies: *Proc. of International Conf. Russ. Science. Shc. Moscow–Voronezh*, 2002, Part 7.

Razd. 1. Section 13. M.: Radio and Communication, 2002, pp. 22–25. [in Russian]

[8] Finoshkina, A.S. Models of plasticity under finite deformations: autoref. diss. on competition of a scientific degree. *Academic Step. Kand. Phys.-Math. Sciences.* Moscow State University. Moscow, 2003, p. 15. [in Russian]

[9] Vasin, R.A., Enikeev, F.U., Kruglov, A.A., and Safiullin, R.V. On the identification of the determining relations by the results of technological experiments. *Proc. of RAS. Mech. of Solid Bodies*, 2003, 2, pp. 111–123. [in Russian]

[10] Belikov, N.V., Zanimonets, Y.M., and Yudin, A.S. Mathematical modeling methods for controlled plastic forming and non-destructive testing of domed shells with a given high-precision load loss of stability. XXXIII sch.-seminar "Mathem. modelir. in probl. of rational nature use". Rostov-on-Don: Publ. of North-Caucasus Science Center of Higher Education, 2005, pp. 133–136. [in Russian]

[11] Yudin, A.S. Modeling of initial imperfections of spherical domes. *Modern. Probl. of Continuum Mech.: Collection of Scientific Works.* Rostov-on-Don: MP "Book", 1995, pp. 210–219. [in Russian]

[12] Yudin, A.S., Kakurin, A.M., and Pyankov, B.G. The critical load of dome-shaped shells under mathematical and physical modeling. *Modern. Probl. of Continuum Mech. Proc. of IV Intern. Science Conf.* Rostov–on–Don: Publ. of North-Caucasus science center of higher education, 1998, 2, pp. 222–225. [in Russian]

[13] Yudin, A.S. and Shepeleva, V.G. Critical pressures of elastic buckling of imperfect spherical shells. *Modern. Probl. of Continuum Mech. Proc. of II Intern. Science Conf.* Rostov-on-Don: MP "Book", 1996, 3, pp. 156–160. [in Russian]

[14] Pyankov, B.G., Kakurin, A.M., and Yudin, A.S. Dome-shaped artificated shells, losing stability. *Modern. Probl. of Continuum Mech. Proc. of IV International Science Conf.* Rostov-on-Don: Publ. of North-Caucasus Science Center of Higher Education, 1998, 2, pp. 129–133. [in Russian]

[15] Pyankov, B.G., Kakurin, A.M., and Yudin, A.S. Experimental and theoretical foundations of of safety membranes artification. Izvestia vuzov. North-Caucasus Region. *Nature Sciences*, 1999, 2, pp. 22–24. [in Russian]

[16] Yudin, A.S. and Yudin, S.A. Modeling of plastic molding of the artificated flapping membrane. *Modern. Probl. of Continuum Mech. Proc. of IV Intern. Science Conf.* Rostov-on-Don: Publ. OOO "CWR", 2006, 1, pp. 290–294. [in Russian]

[17] Yudin, S.A. and Yudin, A.S. Analytics of plastic forming of a spherical dome from a round plate. Mat. models and algor. for imitators. *Physical. Processes: Proc. of Intern. Science-Technical Conf.* Taganrog: Publ. of TGPI, 2006, 1, pp. 212–215. [in Russian]

[18] Yudin, S.A. Plastic deformation of shells of revolution. *Proceedings of the Postgraduate Students of Rostov State University.* Vol. 12. Rostov-on-Don: Publ. "TerraPrint", 2006, pp. 38–40. [in Russian]

[19] Yudin, A.S. and Yudin, S.A. Plastic drawing of the dome from a round plate: Theory and experiment. *Modern. Probl. of Continuum Mech. Proc. of XI Intern. Conf.* Rostov-on-Don: Publ. OOO "CWR", 2007, 1, pp. 255–259. [in Russian]

[20] Yudin, S.A. and Yudin, A.S. Stability of spheroidal shells of variable thickness. *Modern. Probl. of Continuum Mech. Proc. of X Intern. Conf.*, Vol. 1. Rostov-on-Don: MP "Book", 2006, pp. 295–299. [in Russian]

[21] Yudin, A.S. and Yudin, A.S. Conditions of sphericity of the dome in plastic molding of a round plate. *Models and Algorithms for Imitation Phys.-Chemical Processes. Proc. of Intern. Conf.* 8–12 Sept. 2008. Taganrog, Russia. Taganrog: Edit. of TGPI, 2008, pp. 86–94. [in Russian]

[22] Axelrad, E.L. Equations of deformation of shells of rotation and bending of thin-walled rods at large elastic displacements. Izvestia of AS USSR. *Mechanics and mechanical engineering*, 1963, 4, pp. 84–92. [in Russian]

[23] Ambardzumyan, S.A. *General theory of anisotropic shells: monograph.* M.: Nauka, 1974, p. 448. [in Russian]

[24] Akhmerov, A.F. On curves of hardening of materials and their approximation. Izvestia of Higher Educational. *Aviation Equipment.* 1972, 3, pp. 79–86. [in Russian]

[25] Valishvili, N.V. On the limits of applicability of nonlinear equations of flat shells. Izvestia of Higher Educational. *Engineering.* 1972, 5, pp. 14–20. [in Russian]

[26] Vorovich, I.I and Lebedev, L.P. *Functional analysis and its applications in continuum mechanics.* M.: University book, 2000, 320. [in Russian.] Also: Lebedev, L.P., Vorovich, I.I. *Functional analysis in mechanics.* NY: Springer, 2002, p. 249.

[27] Gerlaku, I.D., Morar, V.P., and Shilkrut, D.I. Axisymmetric nonlinear elastic deformations of thin non-linear shells of revolution. *Proc. of VII All-Union Conf. According to the Theory Shells and Plates.* M.: Nauka, 1970, pp. 172–176. [in Russian]

[28] Grigolyuk, E.I., Mamai, V.I., and Frolov, A.N. Study of stability of apology spherical shells on the basis of various equations

of shell theory. *Izvestia of AS USSR. Solid Mechanics.* 1972, 5, pp. 154–165. [in Russian]

[29] Ilyushin, A.A. Plasticity. Part 1. *Elastic-plastic deformation.* M.: Gostekhizdat, 1948, p. 376. [in Russian]

[30] Kamyshev, V.V., Sizova, T.P., Surkin, R.G., and Chernovisov, G.N. The dependence of dome height, thickness and curvature of membranes made of 12X18H10T steel under the change in forming pressure. *Improving the safety of operation of oil refining and petrochemical industries with the help of safety membranes: Collection of scientific papers.* M.: "Neftekhim", 1979, 19, pp. 82–93. [in Russian]

[31] Kachanov, L.M. *Fundamentals of fracture mechanics: monograph.* M.: Nauka, 1974, p. 312. [in Russian]

[32] Kurkin, S.A. *Strength of welded thin-walled vessels working under pressure: monograph.* M: Mashinostroenie, 1976, p. 184. [in Russian]

[33] Lurie, A.I. *Theory of elasticity: monograph.* M.: Nauka, 1970, p. 940. [in Russian.] Also: Lurie, A.I. *Theory of elasticity.* Berlin, Heidelberg: Springer-Verlag, 2005, p. 1036.

[34] Malinin, N.N. *Applied theory of plasticity and creep.* M.: Engineering, 1975, p. 400. [in Russian]

[35] Nadai, A. Theory of flow and fracture of solids. *Eng. transl.* M.: Inostr. Liter., 1954, p. 647. [in Russian]

[36] Pisarenko, G.S. and Mozharovsky, N.S. *Equations and boundary value problems of plasticity and creep theory: Reference manual.* Kiev: Naukova Dumka, 1981, p. 496. [in Russian]

[37] Reissner, E. Finite symmetrical deflections of thin shells of revolution. *Journal of Applied Mechanics.* M.: Mir, 1969, 2, pp. 131–134. [in Russian]

[38] Reissner, E. Finite symmetric deformation of thin shells of rotation. *Journal of Applied Mechanics.* M.: Mir, 1972, 4, pp. 1137–1143. [in Russian]

[39] Hsu, T.C., Shang, H.M., Lee, T.C., and Lee, S.Y. Stresses of plastic flow in the sheet material when forming products of almost spherical shape from it. *Theor. Basics of Eng. Calculations.* M.: Mir, 1975, 97(1), pp. 66–75. [in Russian]

[40] Yudin, A.S. On some nonlinear equations of axisymmetric deformation of shells of rotation. Izvestia of the North-Caucasus Scientific Center of Higher School. *Natural Sciences*, 1973, 4, pp. 93–98. [in Russian]

[41] Reissner, E. On the theory of thin elastic shells. Anniv. vol., *Contrib. Appl. Mech. J. W. Edwards, Ann Arbor, Mich.*, 1949, pp. 231–247.

[42] Reissner, E. On the axisymmetrical deformations of thin shells of revolution. *Proc. 3-rd Symp. Appl. Math.*, 1950, 3, pp. 27–52.

[43] Reissner, E. On the equation for finite symmetrical deflections of thin shells of revolution. *Progr. In Appl. Mech. Prager Anniv.*, 1963, pp. 171–178.

[44] Reissner, E. On finite symmetrical deflections of thin shells of revolution. *Appl. Mech. Trans. of the ASME. Ser. E.*, 1969, 36(2), pp. 267–270.

[45] Yudin, A.S. and Belikov, N.V. Protection of equipment against destruction by overpressure. *Advanced materials — Research and applications*, 2015. Chapter 29, pp. 499–509.

[46] Shchitov, D.V. Behavior of the shell modeling the bottom of the tank. *Proc. of postgraduates students and applicants of Rostov University.* Vol. VII. Rostov-on-Don: Edition of Rostov University, 2001, pp. 17–19. [in Russian]

[47] Shchitov, D.V. Shells of tanks with liquid under static and dynamic effects. Mathematical modeling, computational mechanics and geophysics. *Proceedings of the first school-seminar.* Rostov-on-Don, 14–18 October 2002. Rostov-on-Don: Publishing House "New book", 2002, pp. 170–172. [in Russian]

[48] Shchitov, D.V. Statics and oscillations of shells of revolution containing a fluid. Specialty 01.02.04 — *Mech. of Def. Solid Bodies. Autoabstract for the Degree of Kand. Techn. Sciences*, 2007, p. 142. [in Russian]

[49] Yudin, A.S. *Static and oscillations of shells of rotation with liquid: monograph.* Rostov-on-Don: Publishing House of Southern Federal University, 2014, p. 204. [in Russian]

[50] Zotov, M.V. and Kutasov, I.A. Alignment of buildings and structures using jack systems. *Modern. Probl. of Continuum Mech. Proc. of XIII Intern. Conf.* Vol. I. Rostov-on-Don: Publishing House of Southern Federal University, 2009, pp. 101–105. [in Russian]

[51] Yudin, S.A. and Sigaeva, T.V. Closed axisymmetric shell as a flat jacks. *Advanced Materials. Physics, Mechanics and Applications. Series: Springer Proceedings in Physics*, 2014, 152. Chapter 30, pp. 365–373.

[52] Yudin, A.S., Yudin, S.A., and Sigaeva, T.V. Semi-inversion method in the problem of plastic deformation of cylindrical shell. *Advanced Materials — Studies and Applications*, 2015. Chapter 21, pp. 341–351.

[53] Yudin A.S. Plastic forming model for axisymmetric shells. *Materials Physics and Mechanics*, 2018, 37(2), pp. 176–183.

[54] Yudin, A.S. and Shchitov, D.V. Reduction of stresses in the shell by the plastic change of its form. *Advanced Materials — Studies and Applications.* Nova Science Publishers, 2019. Chapter 24.

[55] Yudin, A.S. Simulation of the work of Flat Jack in Large Displacements. *Advanced Materials — Studies and Applications. —* Nova Science Publishers, 2019. Chapter 41.

[56] Vodyanik, V.I., Malakhov, N.N., Poltavsky, V.T., and Shelyuk, I.P. Safety Membranes: Reference. Manual. *M.: Chemistry*, 1982, p. 144. [in Russian]

[57] Glinskikh, A. World market of CAD / CAM / CAE systems. www.ca tia.spb.ru. [in Russian]

[58] Plis, A.I. and Slivina, N.A. *Mathcad: mathematical workshop*. M.: Finance and Statistics, 1999, p. 656. [in Russian]

[59] Makarov, E. *Engineering calculations in MathCad. Training Course*. SPb: Peter, 2003, p. 448. [in Russian]

[60] Zotov, M.V. and Kutasov, I.A. Equalization of buildings and structures using Jack systems. *Modern problems of continuum mechanics. Proc. of the VIII Intern. conf. Rostov-on-Don*. Rostov-on-Don: Publishing House of Southern Federal University, 2009, 1, pp. 101–105. [in Russian]

[61] Yudin, A.S. Effective models for composite shells of rotation. Izvestiya of North-Caucasus Scientific Center of Higher Education. *Natural Sciences*, 2000, 3, pp. 184–188. [in Russian]

[62] Yudin, A.S., Yudin, S.A., and Kutasov, I.A. "Tor-plate" type rotation shell at large displacements. *Modern problems of continuum mechanics. Proc. of the XIII Intern. Conf.* Publishing House of Southern Federal University, 2009, 2, pp. 201–205. [in Russian]

[63] Bronstein, I.N. and Semendyaev, K.A. *Handbook of mathematics for engineers and university students*. M.: Nauka, 1980, p. 976. [in Russian]

[64] Novozhilov, V.V. *Theory of thin shells: monograph*. L.: Sudpromgiz, 1962, p. 431. [in Russian]

[65] Karmishin, A.V., Lyaskovets, V.A., Myachenkov, V.I., and Frolov, A.N. *Statics and dynamics of thin-walled shell structures*. M.: Mashinostroenie, 1975, p. 376. [in Russian]

[66] Myachenkov, V.I. and Grigoriev, I.V. *Calculation of composite shell structures on a computer: Reference guide*. M.: Mashinostroenie, 1981, p. 216. [in Russian]

[67] Yudin, A.S. and Shchitov, D.V. Analysis of shells with discontinuous parameters. Izv. Vuzov. North Caucasus region. *Natural Sciences*. Appendix, 2004, 3, pp. 22–29. [in Russian]

Index

Printed in the United States
by Baker & Taylor Publisher Services